T0192428

SpringerBriefs in Applied Sciences and Technology

Nanotheranostics

Series Editors

Subramanian Tamil Selvan, Institute of Materials Research & Engineering, National University of Singapore, Singapore, Singapore

Karthikeyan Narayanan, Singapore, Singapore

Padmanabhan Parasuraman, Singapore, Singapore

Paulmurugan Ramasamy, Stanford University School of Medicine, Palo Alto, CA, USA

More information about this subseries at http://www.springer.com/series/13040

Devasena T.

Nanotechnology-COVID-19 Interface

Springer

Devasena T.
Centre for Nanoscience and Technology
Anna University
Chennai, Tamil Nadu, India

ISSN 2191-530X   ISSN 2191-5318 (electronic)
SpringerBriefs in Applied Sciences and Technology
ISSN 2197-6740   ISSN 2197-6759 (electronic)
Nanotheranostics
ISBN 978-981-33-6299-4   ISBN 978-981-33-6300-7 (eBook)
https://doi.org/10.1007/978-981-33-6300-7

This Springer imprint is published by the registered company Springer Nature Singapore Pte Ltd.
The registered company address is: 152 Beach Road, #21-01/04 Gateway East, Singapore 189721,
Singapore

Contents

Chapter 1
Introduction to COVID-19

1.1 Coronavirus

Corona means "crown" in Latin. Coronaviruses are group of viruses characterized by crown-like projections of glycoproteins on their surface. According to the International Committee on the Taxonomy of Viruses, coronaviruses come under a broad order called Nidovirales. Nidovirales is an order of enveloped, positive-strand RNA viruses which infect mammals, birds, reptiles, amphibians, fish, arthropods, mollusks, and helminths. Nidovirales order encompasses four families, namely Coronaviridae, Arteriviridae, Roniviridae and Mesoniviridae. Among those four families, the Coronaviridae family is considered here. Coronaviridae is divided into two subfamilies Coronavirinae and Toronavirinae. The Coronavirinae subfamily is further divided into four genus, namely Alphacoronavirus (αCoV), Betacoronavirus (βCoV), Gammacoronavirus (γCoV) and Deltacoronavirus. Alpha and Betacoronaviruses typically infect mammals only, while the Gamma and Deltacoronaviruses typically infect avian species and sometimes mammals (Cui et al. 2019). The pandemic viruses, namely severe acute respiratory syndrome coronavirus (SARS CoV), the middle east respiratory syndrome coronavirus (MERS CoV) and the severe acute respiratory syndrome coronavirus-2 or the novel coronavirus 2019 (SARS CoV2 / the nCOV19) are classified under the genus Betacoronavirus (Chan et al. 2013). This taxonomy is schematized in Fig. 1.1.

1.2 Severe Acute Respiratory Syndrome Coronavirus (SARS CoV)

The severe acute respiratory syndrome (SARS) outbreak caused by the SARS CoV was first identified in the Guangdong Province of Southern China during 2002–2003, which killed approximately 900 people. The pandemic spread rapidly to around 25 countries with flu-like symptoms including malaise, myalgia, headache, diarrhea,

© The Author(s), under exclusive license to Springer Nature Singapore Pte Ltd. 2021   1
Devasena T., *Nanotechnology-COVID-19 Interface*, Nanotheranostics,
https://doi.org/10.1007/978-981-33-6300-7_1

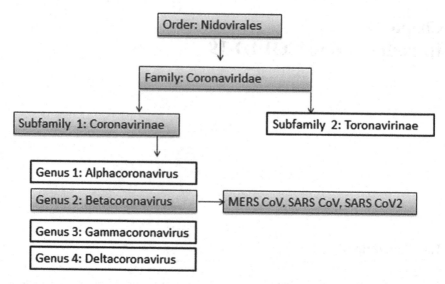

Fig. 1.1 Taxonomical hierarchy of coronaviruses. The way of SARS CoV2 is shaded. (*Legend: The SARS CoV2 belongs to the order Nidovirales, family Coronaviridae, subfamily Coronavirinae and genus Betacoronavirus*)

shivering (rigors), fever, headache and respiratory problems such as cough and shortness of breath. SARS CoV was thought to be an animal virus, perhaps of bat origin, that spread to other animals (civet cats) and first infected humans in 2002.

SARS CoV transmission was peaked during second week after infection. However, in countries like Toronto in Canada, Hong Kong, China, Chinese Taipei, Singapore and Hanoi, there was a considerable transmission while China was considered as the potential zone of SARS CoV2 reemergence. In this peak period, viral titer was reported to be higher in respiratory secretions leading to clinical deterioration. After 2003, no areas of the world reported natural transmission of SARS. But, SARS was reported to spread again from laboratory accidents in Singapore and Chinese Taipei and once in southern China where the source of infection is anonymous. However, appropriate non-pharmaceutical intervention played a vital role in ending that global outbreak. Later, the SARS CoV was classified as a member of A-lineage beta coronavirus, and the genome was sequenced (Rota et al. 2003).

1.3   Middle East Respiratory Syndrome Coronavirus (MERS CoV)

After a decade from the outbreak of SARS pandemic, the middle east respiratory syndrome (MERS) outbreak of MERS CoV origin occurred in Saudi Arabia during 2012, which was found to be prevalent in middle east travelers. MERS CoV produced same respiratory symptoms but less severe than the SARS CoV and also capable of causing renal failure. MERS CoV belongs to Betacorona virus group of the C-lineage category. MERS is accompanied by metabolic syndromes such as diabetes mellitus, cardiovascular diseases and obesity, which can subsequently interfere with innate and humoral immunity and can render patients more susceptible to infectious diseases. By 2012, researches and scientists focused on different aspects of coronaviruses such as structure, proteomics, genomics, exploring the life cycle, infective mechanisms, anticoronavirals, drug development, diagnostic tools and management strategies (Vincent et al. 2007; Savarino et al. 2006).

1.4   Severe Acute Respiratory Syndrome Coronavirus-2 (SARS CoV2)

In December 2019, the World Health Organization (WHO) alerted the entire world about the cases of pneumonia of unknown cause in Wuhan City, Hubei Province, China. This disease has rapidly increased the laboratory-confirmed infections causing 80 deaths within a month of its outbreak (Zhou et al. 2020). The disease had become pandemic and transmitted to more than 203 countries and territories, including community transmission in countries like the USA, Germany, France, Spain, Japan, Singapore, South Korea, Iran, India and Italy. The causative agent of this pandemic was initially called as a novel coronavirus of 2019 (nCoV-19) because of the clinical symptoms like fever, malaise, dry cough and dyspnea, severe lower respiratory tract infection with acute respiratory distress syndrome which are similar to the symptoms of SARS infection. Extrapulmonary manifestations such as diarrhea, lymphopenia, deranged liver, abnormal renal functions and multiorgan dysfunction also occurred in both immunocompetent and immunocompromised host cells (Peiris et al. 2003; Yeung et al. 2016). The pandemic was severe in persons with travel history to Wuhan (Chan et al. 2020). The ICTV termed the nCoV-19 as the SARS CoV2 which belonged to the Betacoronavirus A-lineage and showed many similarities with the structure and symptoms of SARS CoV. According to the cryo electron microscopic studies, the ultrastructure of SARS CoV2 shows spherical or elliptical morphology with 70–90 nm diameter (Kim et al. 2020). After naming the causative agent, the WHO has announced the official name of the pandemic as the Corona Virus Disease 2019 or COVID-19 (WHO 2020).

1.5   Origin of SARS CoV2

SARS CoV2 originated in bats and believed to be transmitted first to humans in an open wet market of China. The entire world consequently realized the SARS

CoV2 pandemic as a natural evolution. After a natural selection process in a non-human host, SARS CoV2 evolved as a pathogenic organism and then transmitted to human host. Even the previous outbreaks of SARS and MERS occurred in the same way, where the humans hosted the virus after exposure to the non-human hosts civets and camels, respectively. Though there is no strong evidence for bat-to-human transmission of SARS CoV2, the genome of the later resembles bat coronavirus. It was documented that an intermediate amplifying host may be involved between bat and human. Pangolins, traded illegally in Asia were thought to be a potential amplifying intermediate host by some studies (Lam et al. 2020; Zhang et al. 2020). Civet cats, swine, cats, ferrets, non-human primates (NHPs) were also reported to be the intermediates as they have the receptors for SARS CoV2 (Letko et al. 2020; Wan et al. 2020).

There are two possible origin of SARS CoV2: (i) natural selection in an animal host before zoonotic transfer and (ii) natural selection in humans following zoonotic transfer (Zheng 2020). Various aspects of SARS CoV, MERS CoV and SARS CoV2 have been discussed by several researchers (Rabaan et al. 2020) and are compared in Table 1.1. Evolution of coronaviruses is shown in Fig. 1.2. Humans of any age can get infected by SARS CoV2, but adults of middle age and elderly are most commonly affected. The genome of the SARS CoV2 was isolated from clinical samples of patients who tested positive for SARS CoV2 in Wuhan (Chan et al. 2020b). Analysis of genome sequence of SARS CoV2 showed much similarity with SARS CoV genome, confirmed that SARS CoV and SARS CoV2 belong to the same family and both of them have natural reservoir as origin, may be bat (Zhou et al. 2020). SARS CoV2 can be stable in air for two to three hours and on surfaces of plastic, stainless steel and metal (Doremalen et al. 2020).

1.6   Geographical Distribution of SARS CoV2

The geographical incidence and the distribution of COVID-19 vary due to region-dependent evolution and variation in the genome mutation pattern. The mutation frequency varies in the viral strains of different geographical regions (Table 1.2). The mutation frequency was the lowest in the Asian region and highest in the North American region as: Asia < Oceania < Europe < North America. Asian strains show higher mutation in non-structural protein 4 (nsp4) and open reading frame protein 8a (Refer Chap. 2, for details on this genes and proteins). Oceania region samples possess high mutations in nsp6 and nsp2.

Interestingly, European country samples are mutated in the genes encoding spike protein and RNA-dependent RNA polymerase (RdRp) while North American samples are highly mutated in the genes coding for RdRp and papain like protease (PLpro). The significance of the protein products of these genes are discussed in Chaps. 2 and 5. Thus, the incidence, the severity and the recovery rate of COVID-19 are not constant, and it varies among different geographical regions. SARS CoV2 death also varies, among different geographical areas. For example, fatality rate is

Table 1.1 Comparison of the coronaviruses

Details	SARS CoV	MERS CoV	SARS CoV2 (nCoV-19)	References
Outbreak	November 2002	April 2012	December 2019	Zhu et al. (2020)
Geographical location of the first outbreak	Guangdong, China	Saudi Arabia	Wuhan, China	Willman et al. (2019), Zeng et al. (2020), Wu et al. (2020)
Natural reservoir	Bat	Bat	Bat	Wang et al. (2006), Mohd et al. (2016), Hossain et al. (2020)
Intermediate transmitters to humans	Via civet cats and raccoon dogs	Via dromedary camel	The intermediate host may be pangolin Not identified and confirmed	Wang and Eaton (2007); Omrani et al. (2015); Zhang et al. (2020)
Relative Rate of transmission	High	High	Very rapid	Liu et al. (2020)
Incubation period	2–7	5–6	7–14	Zhu et al. (2020)
Affinity of virus glycoprotein antigen to host receptor	Strong	Strong	Strongest	Teng et al. (2020)
Prominent host receptor for entry	Angiotensin converting enzyme-2 (ACE2)	Dipeptidyl peptidase-4 (DPP4), CD26	Angiotensin converting enzyme-2 (ACE2) and TMPRSS2 protease for priming	Li (2013), Li et al. (2020), Hoffmann et al. (2020)
Mortality rate of the infection	Medium (9.6%)	Higher (35%)	Low (3.4%)	Abdelrahman et al. (2020)
Clinical symptoms: fever, cough, headache, malaise	Positive	Positive	Positive	Zhu et al. (2020)

three times higher out of China as compared to the origin China (Pachetti et al. 2020). However, several variable factors which are different among countries like management strategies people movement restrictions, isolation and quarantine, different immunity level, social distancing and different genetic population, herd immunity also influence the geographical incidence, the mortality rate and the degree of decline in the infection rate.

COVID-19 pandemic has become an uncontrollable global issue as there is no specific clinically approved drugs and vaccines marketed so far. The patients are

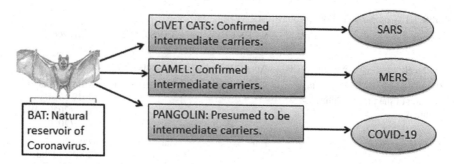

Fig. 1.2 Evolution of coronaviruses from natural reservoir to humans through non-human intermediate hosts

Region	Country	Relative Mutation frequency	Remarkable mutation spot
Asia	China Japan Myanmar Nepal India	Low	Nsp4 ORF8a
Oceania	Australia	High	Nsp6 and nsp2
Europe	Spain Portugal UK Netherlands Italy Germany Switzerland France Luxemburg Sweden Finland Denmark Belgium	Very high	Spike protein RdRp
North America	USA Canada	Very high	RdRp Papain-like protease (nsp3)

Table 1.2 Geographical variation in COVID-19 depending upon the SARS CoV2 genomic mutation (Pachetti et al. 2020)

however being managed by the drugs already available in the market but are capable of alleviating the symptoms of COVID-19. In this delicate scenario, it is essential for the pharmaceutical companies, researchers and scientists to clearly elucidate the following aspects of SARS CoV2 to successfully design the drugs, vaccines and diagnosis kits to fight COVID-19:

• The structure and virucidal epitopes of SARS CoV2

- The proteomics of SARS CoV2
- The genomic pattern of SARS CoV2
- The mechanism involved in each stages of SARS CoV2 life cycle
- Existing and promising methods of corona viral targeting and detection.

In this context, the unique properties of nanomaterial and advancements in nanotechnology pertaining to diagnosis of specific proteins, DNA sequences and single molecules, targeted and sustained delivery of drugs, nanoparticulate vaccines, the innate therapeutic efficacy of some nanoparticles per se (Refer Chaps. 4 through 7) should be considered by the researchers for management of COVID-19. There are numerous potential nanotechnology-based research targets for management of SARS CoV2 infection (Chan 2020), and at the same time the publications and patents in this area are scarce. (Uskokovic 2007; Kostarelos 2020).

Therefore, the subsequent chapters of this book systematically address various key aspects of SARS CoV2, the interface between nanotechnology and SARS CoV2/COVID-19, the existing and promising nanotheranostic and vaccine tools. The chapters deserve significance as they can open up novel avenues for the repurposing, designing and marketing of nanotechnology-based management tools to combat COVID-19.

References

Abdelrahman Z, Li M, Wang X. Comparative review of SARS-CoV-2, SARS-CoV, MERS-CoV, and influenza a respiratory viruses. Front Immunol. 2020;11:552909.

Chan JF, Kok KH, Zhu Z, Chu H, To KK, Yuan S, Yuen KY. Genomic characterization of the 2019 novel human-pathogenic coronavirus isolated from a patient with atypical pneumonia after visiting Wuhan. Emerg Microb Infect. 2020;9(1):221–36.

Chan JF, To KK, Tse H, Jin DY, Yuen KY. Interspecies transmission and emergence of novel viruses: lessons from bats and birds. Trends Microbiol. 2013;21(10):544–55.

Chan JF, Yuan S, Kok KH, Wang To KK, Chu H, Yang J, Xing F, Liu J, Yan Yip CC, Shan Poon RW, Tsoi HW, Fai Lo SK, Chan KH, Poon VK, Chan WM, Daniel J, Cai JP, Cheng VCC, Chen H, Hui CKM. A familial cluster of pneumonia associated with the 2019 novel coronavirus indicating person-to-person transmission: a study of a family cluster. Lancet. 2020;395(10233):514–23.

Chan WCW. Nano research for COVID-19. ACS Nano. 2020;14(4):3719–20.

Cui F, Li Z, Shi L. Origin and evolution of pathogenic coronaviruses. Nat Rev Microbiol. 2019;17:181–92.

Doremalen NV, Morris DH, Holbrook MG, Gamble A, Williamson BN, Tamin A, Harcourt JL, Thornburg NJ, Gerber SI, Lloyd-Smith JO, Wit ED, Munster VJ. Aerosol and surface stability of SARS-CoV-2 as compared with SARS-CoV-1. N Engl J Med. 2020;382:1564–7.

Hoffmann M, H Kleine-Weber, Schroeder S, Krüger N, Herrler T, Erichsen S, Schiergens TS, Herrler G, Wu NH, Nitsche A, Müller MA, Drosten C, Pöhlmann S. Sars-cov-2 cell entry depends on ace2 and tmprss2 and is blocked by a clinically proven protease inhibitor. Cell. 2020;181(2):271-280.e8.

Hossain MG, Javed A, Akter S, Saha S. SARS-CoV-2 host diversity: an update of natural infections and experimental evidence. J Microbiol Immunol Infect. 2020;S1684–1182(20):30147-X. Advance online publication.

Kim JM, Chung YS, Jo HJ, Lee NJ, Kim MS, Woo SH, Park S, Kim JW, Kim HM, Han MG. Identification of coronavirus isolated from a patient in Korea with COVID-19. Osong Publ Health Res Perspect. 2020;11:3–7.

Kostarelos K. Nanoscale nights of COVID-19. Nat Nanotechnol. 2020;15:343–4.

Lam TTY, Shum MHH, Zhu HC, Tong YG, Ni XB, Liao YS, Wei W, Cheung WYM, Li WJ, Li LF, Leung GM, Holmes EC, Hu YL, Guan Y. Identification of 2019-nCoV related coronaviruses in Malayan pangolins in southern China. BioRxiv Preprint. 2020. https://doi.org/10.1101/2020.02.13.945485.

Letko M, Marzi A, Munster V. Functional assessment of cell entry and receptor usage for functional assessment of cell entry and receptor usage for SARS-CoV-2 and other lineage betacoronaviruses. Nat Microbiol. 2020;5:562–569.

Li Y, Zhang Z, Yang L, Lian X, Xie Y, Li S, Xin S, Cao P, Lu J. The MERS-CoV receptor DPP4 as a candidate binding target of the SARS-CoV-2 spike. iScience 2020;23(6):101160.

Li F. Receptor recognition and cross-species infections of SARS coronavirus. Antiviral Res. 2013;100:246–54.

Liu J, Xie W, Wang Y, Xiong Y, Chen S, Han J, Wu Q. A comparative overview of COVID-19, MERS and SARS: Review article. Int J Surg. 2020;81:1–8.

Mohd HA, JA Al-Tawfiq, Memish ZA. Middle East respiratory syndrome coronavirus (MERS-CoV) origin and animal reservoir. Virol J. 2016;13:87.

Omrani AS, Al-Tawfiq JA, Memish ZA. Middle East respiratory syndrome coronavirus (MERS-CoV): animal to human interaction. Pathog Glob Health. 2015;109(8):354–62.

Pachetti M, Marini B, Benedetti F, Giudici F, Mauro E, Storici P, Masciovecchio C, Angeletti S, Ciccozzi M, Gallo RC, Zella D, Ippodrino R. Emerging SARS-CoV-2 mutation hot spots include a novel RNA-dependent-RNA polymerase variant. J Trans Med. 2020;18(1):179.

Peiris JSM, Lai ST, Poon LLM, Guan Y, Yam LYC, Lim W, Nicholls J, Yee WKS, Yan WW, Cheung MT, Cheng VCC, Chan KH, Tsang DNC, Yung RWH, Ng TK, Yuen KY. coronavirus as a possible cause of severe acute respiratory syndrome. Lancet. 2003;361(9366):1319–25.

Rabaan AA, Al-Ahmed SH, Haque S, Sah R. SARS-CoV-2, SARS-CoV, and MERS-COV: a comparative overview. InfezMed. 2020;28(2):174–84.

Rota PA, Oberste MS, Monroe SS, Nix WA, Campagnoli R, Icenogle JP, Peñaranda S, Bankamp B, Maher M, Chen MH, Tong S, Tamin A, Lowe L, Frace M, De Risi JL, Chen Q, Wang D, Erdman DD, Peret TCT, Burns C, Ksiazek TG, Rollin PE, Sanchez A, Liffick S, Holloway B, Limor J, McCaustland K, Olsen-Rasmussen M, Fouchier R, Günther S, Osterhaus ADME, Drosten C, Pallansch MA, Anderson LJ, Bellini WJ. Characterization of a novel coronavirus associated with severe acute respiratory syndrome. Science. 2003;300:1394–9.

Savarino A, Buonavoglia C, Norelli S, Trani LD, Cassone A. Potential therapies for coronaviruses. Expert Opin Ther Pat. 2006;16(9):1269–88.

Teng S, Sobitan A, Rhoades R, Liu D, Tang Q, Systemic effects of missense mutations on SARS-CoV-2 spike glycoprotein stability and receptor-binding affinity. Briefings Bioinf. 2020;bbaa233:1–15.

Uskokovic V. Nanotechnologies: what we do not know. Technol Soc. 2007;29(1):43–61.

Vincent C. Cheng C, Susanna K. Lau P, Patrick C. Woo, Yuen K Y. Severe acute respiratory syndrome coronavirus as an agent of emerging and reemerging infection. Clin Microbiol Rev. 2007;20:660–694.

Wan Y, Shang J, Graham R, Baric RS, Li F. Receptor recognition by novel coronavirus from Wuhan: An analysis based on decade-long structural studies of SARS. J Virol. 2020;94:e00127-e220.

Wang LF, Eaton BT. Bats, civets and the emergence of SARS. Curr Top Microbiol Immunol. 2007;315:325–44.

Wang LF, Shi Z, Zhang S, Field H, Daszak P, Eaton BT. Review of bats and SARS. Emerg Infect Dis. 2006;12(12):1834–40.

WHO. WHO Director-General's remarks at the media briefing on 2019-nCoV on 11 February 2020. https://www.who.int/dg/speeches/detail/who-director-general-s-remarks-at-the-media-briefing-on-2019-ncov-on-11-february-2020.

Willman M, Kobasa D, Kindrachuk J. A comparative analysis of factors influencing two outbreaks of middle eastern respiratory syndrome (MERS) in Saudi Arabia and South Korea. Viruses. 2019;11(12):1119.

Wu P, Hao X, EHY L, Wong JY, KSM L, Wu JT, Cowling BJ, Leung GM. Real-time tentative assessment of the epidemiological characteristics of novel coronavirus infections in Wuhan, China, as at 22 January 2020. Eurosurveillance 2020;25:2000044.

Yeung ML, Yao Y, Jia L, Chan JF, Chan KH, Cheung KF, Chen H, Poon VK, Tsang AK, To KK, Yiu MK, Teng JL, Chu H, Zhou J, Zhang Q, Deng W, Lau SK, Lau JY, Woo PC, Chan TM, Yung S, Zheng BJ, Jin DY, Mathieson PW, Qin C, Yuen KY. MERS coronavirus induces apoptosis in kidney and lung by upregulating Smad7 and FGF2. Nat Microbiol. 2016;1(3):16004.

Zeng ZQ, Chen DH, Tan WP, Qiu SY, Xu D, Liang HX, Chen MX, Li X, Lin ZS, Liu WK, Zhou R. Epidemiology and clinical characteristics of human coronaviruses OC43, 229E, NL63, and HKU1: a study of hospitalized children with acute respiratory tract infection in Guangzhou, China. Eur J Clin Microbiol Infect Dis. 2018;37(2):363–9.

Zeng ZQ, Chen DH, Tan WP, Qiu SY, Xu D, Liang HX, Chen MX, Li X, Lin ZS, Liu WK, Zhou R, Zhang T, Qunfu W, Zhang Z. Probable pangolin origin of SARS-CoV-2 associated with the COVID-19 outbreak. Curr Biol. 2020;30(8):1578.

Zhang T, Wu Q, Zhang Z. Pangolin homology associated with 2019-nCoV. BioRxiv Preprint. 2020. https://doi.org/10.1101/2020.02.19.950253.

Zheng J. SARS-CoV-2: an emerging coronavirus that causes a global threat. Int J Biol Sci. 2020;16(10):1678–85.

Zhou P, Yang XL, Wang XG, Hu B, Zhang L, Zhang W, Si HR, Zhu Y, Li B, Huang CL, Chen HD, Chen J, Luo Y, Guo H, Jiang RD, Liu MQ, Chen Y, Shen XR, Wang X, Zheng XS, Zhao K, Chen QJ, Deng F, Liu LL, Yan B, Zhan FX, Wang YY, Xiao GF, Shi ZL. A pneumonia outbreak associated with a new coronavirus of probable bat origin. Nature. 2020;579:270–3.

Zhu Z, Lian X, Su X, et al. From SARS and MERS to COVID-19: a brief summary and comparison of severe acute respiratory infections caused by three highly pathogenic human coronaviruses. Respir Res. 2020;21:224.

Chapter 2
Structure of SARS CoV2

SARS CoV2 is an enveloped RNA virus ranging in diameter approximately between 70 and 90 nm (Kim et al. 2020a, b). Thus, SARS CoV2 may be referred as a viral nanostructure (Refer Chap. 6; chloroquine section). The protein envelope forms the outer structure which encloses the inner structure and the RNA genome-protein complex (Zhang et al. 2020a, b).

2.1 The Structural Proteins

The protein envelope is made of four structural proteins. They are the spike (S) glycoprotein, the envelope (E) protein, the membrane (M) protein and the nucleo-capsid (N) protein (Shereen et al. 2020). The structure of SARS CoV2 is illustrated in Fig. 2.1.

2.1.1 The Spike Glycoprotein

The S protein is a class I fusion protein project outward giving a crown-like appearance. It is a transmembrane protein with two subunits S1 and S2. A furin cleavage site is present at the S1-S2 interface. The S protein of SARS CoV2 shows 76% and 29% homology with the counterparts of SARS CoV and MERS CoV, respectively. The S1 subunit contains the receptor-binding domain (RBD) which is mainly involved in the attachment of virus to the host receptor (ACE2) and penetration of the virus into the host cell (Refer Chap. 3). Thus, S protein is significant in antigenicity and host attachment which is the very first and the important stage of the viral life cycle. The natural affinity of the SARS CoV2 spike protein to human ACE2 is probably the consequence of natural selection on a human or human-like ACE2. This is a solid proof that SARS CoV2 is not the product of purposeful genetic engineering

© The Author(s), under exclusive license to Springer Nature Singapore Pte Ltd. 2021
Devasena T., *Nanotechnology-COVID-19 Interface*, Nanotheranostics,
https://doi.org/10.1007/978-981-33-6300-7_2

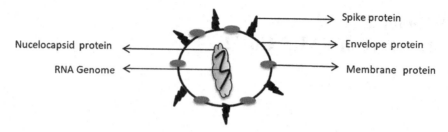

Fig. 2.1 Structural features of SARS CoV2

of laboratory origin. The RBD of SARS CoV2 virus is very much similar to that of pangolin suggesting the later to be an intermediate carrier host. The S protein in SARS CoV2 is longer as compared to the S proteins of the bat coronaviruses, the SARS CoV and the MERS CoV. The S2 subunit contains conserved fusion peptide (FP), heptad repeat (HR), transmembrane domain (TM) and cytoplasmic domain (CP). The genome encoding the SARS CoV2 S2 subunit exhibits 93% similarity with its counterpart in bat coronaviruses, which confirms bat as a natural reservoir. S2 subunit enables fusion and entry of virus during host transmission. The furin cleavage site is a polybasic site which is the target for cleavage by furin protease. This S1-S2 interface is responsible for priming, cell fusion and high host infectivity. The second feature that is unique to SARS CoV2 which are not present in other Betacoronaviruses is this polybasic cleavage site with proline residues and the three adjacent predicted O-linked glycans. The S1-S2 junction is susceptible to mutation, suggesting a natural evolutionary process. The exact function of O-linked glycan is yet to be elucidated, but it is considered as a mucin-like domain capable of shielding the S protein epitopes as a strategy for immunoevasion (Andersen et al. 2020). Structural information on spike protein of SARS CoV2 reveals that cross-neutralizing antibodies against the conserved S epitopes (antigens) can be elicited upon vaccination (Walls et al. 2020).

2.1.2   The Envelope (E) Protein

The E protein of SARS CoV2 is the shortest structural protein and is of transmembrane in location (8–12 kDa, 75–109 amino acid residues). E protein shows 94% and 34% sequence similarity with that of SARS CoV and MERS CoV, respectively. SARS CoV2 E protein is involved in the viral assembly, maturation, budding and virulence. The E protein consists of an ectodomain with N terminus and an endodomain with the C terminus both hydrophilic in nature and separated by a larger trans membrane domain. The C terminus contains proline-rich beta coil motifs which functions as golgi complex-targeting signal responsible for the viral assembly and maturation. During the life cycle of SARS CoV2, the E protein is synthesized in the host cell. Surprisingly, only a small quantity of E protein is used for the structural framework of the virions. The major quantity of the protein however gets stuck in the intracellular trafficking at the subcellular compartments like endoplasmic reticulum, golgi apparatus and ER-golgi intermediate compartment. These proteins get bypassed to the process of viral assembly and maturation. The C terminus has a zonula occludens-1 protein-binding domain (also called PDZ-binding domain). PDZ-binding domain is a major determinant of viral virulence and pathogenesis as it is involved in the protein–protein interaction and infection. The bigger transmembrane domain that spans the membrane is hydrophobic as it is rich in valine and leucine residues. The transmembrane domain contains alpha helical motifs which forms the ion channels. Ion channel activity is the significant feature of E protein, which is responsible for the pathogenesis of SARS CoV2. The functions of E protein can be confirmed by E protein-negative recombinant coronavirus. The recombinant virus resulted in crippled viral maturation, incompetent progeny and lesser viral titer after infecting the host (Jimenez-Guardeno et al. 2014). The PDZ-binding domain can be attenuated into a promising vaccine candidate to fight COVID-19.

2.1.3 The Membrane (M) Protein

The M protein of SARS CoV2 is 25–30 kDa with an N terminal endodomain, C terminal exodomain and a transmembrane helical domain. It shows 98% sequence homology with its counterpart from bat and pangolin isolates. The M protein shows 90% and 39% sequence identity as compared with the M protein of SARS CoV and MERS CoV, respectively. However, the SARS CoV2 M protein shows a serine residue insertion at the N terminus which makes it unique as this position is vacant in bat M protein and replaced by asparagine in pangolin M protein. The M protein interacts with all other structural proteins, stabilizes them and enables the organization of viral envelope. Therefore, it is also called as the central organizer of viral assembly. For example, M protein stabilizes the N proteins-RNA complex inside the virion. Also, M protein interacts with the S protein and enables its assembly into new virions (Schoeman and Fielding 2019; Bianchi et al. 2020).

2.1.4 The Nucleocapsid (N) Protein

The N protein has a C terminal and a N terminal domain and shows 90% homology with the SARS CoV and 29% homology with the MERS CoV counterparts. N protein is highly phosphorylated, and so it has high affinity for the positive RNA of the virus, forming the nucleocapsid. Thus, the main function of N protein is packaging of RNA genome by binding through both the C and the N terminus. This function however depends on RNA packaging signals which is a prerequisite for initiating the viral RNA packaging (Narayanan and Makino 2001). In addition to the RNA, N protein is also distributed in the endoplasmic reticulum of the infected host. Hence, the other function of N protein is the host response mechanism against the infection. As the N

Table 2.1 Percentage sequence homology for SARS CoV and MERS CoV

Protein	SARS CoV (%)	MERS CoV
S protein	76	29
E protein	94	34
M protein	90	39
N protein	90	50

Table 2.2 Functions of SARS CoV2 structural proteins

Structural proteins	Functions
Spike protein, S1 subunit	Binding of virus envelope with host receptor
Spike protein, S2 subunit	Entry into host cell
Spike protein, furin cleavage site	Priming, cell–cell fusion and infectivity
Spike protein, mucin-like domain	Predicted to be involved in epitope shielding and immunoevasion
Envelope (E) protein	Assembly, maturation and pathogenesis of viruses
Membrane (M) protein	Organizing structural proteins to form viral envelope and binding to form a stable N protein-RNA complex
Nucleocapsid (N) protein	Involved in RNA packaging

protein has affinity for RNA, the former can be used in the detection of COVID-19 using RNA sequence as targets. Further, N proteins are capable of eliciting IgG, IgA and IgM antibodies in the host, suggesting that N protein is antigenic marker. N protein is interferon antagonistic and supports viral replication (Schoeman and Fielding 2019; Tai et al. 2020).

Altogether, the structural proteins of COVID-19 virus share high percentage of sequence identity with the counterparts of SARS CoV proteins and comparatively low identity with the MERS CoV counterparts (Table 2.1). The proteins contribute to several complex functions with reference to the virus and the host (Astuti and Ysrafil 2020; Jiang et al. 2020). The functions of all the structural proteins discussed above are summarized in Table 2.2.

2.2   The RNA Genome of SARS CoV2

Mapping studies using full-length genome sourced from SARS CoV2-infected patients reveal several features (Table 2.3), and this information would be useful for better elucidation of structure, functions, antiviral targets, vaccine targets and diagnostic marker identification. The genome structure of SARS CoV2 is schematized in Fig. 2.2. The genome has 96.2%, 79% and 50% identity with that of bat coronavirus, SARS CoV and MERS CoV, respectively. The genome is a positive linear strand molecule with the leader RNA at the 5′ end. The SARS CoV2 genome

Table 2.3 Features of SARS CoV2 genome

Genome	Features
Nature	Positive single-stranded genome
Size	30 Kb
	Largest RNA viral genome known
Canonical RNA genome:	
Open reading frames (ORFs)	14 numbers
Product of 5′ ORFs (ORF1a and ORF1b)	Polyproteins which are cleaved into sixteen non-structural proteins (nsp1 to nsp16) which forms the replicase transcriptase complex-RTC)
	Subgenomic mRNAs (sgmRNAs)
Proteolytic products of PP	Papain-like protease
	Chymotrypsin-like protease
	Ortho methyl transferase
	RNA-dependent RNA polymerase
	Nucleoside triphosphatase, helicase
Subgenomic mRNA products	Four structural proteins (S, E, M and N proteins)
	Putative accessory proteins
Hemagglutinin-esterase gene	Absent
Mutation hotspot	RdRp gene
Highest gene expression site	Lungs

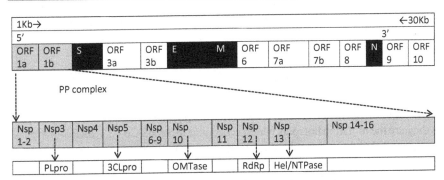

Fig. 2.2 Genome structure of SARS CoV2. (Gray shade represents genes encoding non-structural proteins or nsps (functional proteins). Black shade represents genes encoding structural proteins. ORF—open reading frames; S—spike protein gene; E—envelope protein gene; M—membrane protein gene, N—nucleocapsid protein gene. PP—polyprotein complex formed from ORF1ab which undergoes proteolytic cleavage. PLpro-papain like protease; 3CLpro-chymotrypsin like protease; OMTase-orthomethyltransferase; RdRp-RNA-dependent RNA polymerase; Hel-helicase; NTPase-nucleoside triphosphatase)

possesses 30 kilobases, and it is the largest RNA viral genome known. (Zhou et al. 2020; Chan et al. 2020; Wu et al. 2020) *SARS CoV genome is devoid of hemagglutinin esterase gene which is naturally encoded in the A-lineage Betacoronavirus* (Chan et al. 2020).

2.2.1   The Open Reading Frames (ORFs)

The SARS CoV2 genomic RNA contains 14 open reading frames (ORFs) flanked by a 5′ cap structure and a 3′ poly-A tail. Two-third of the genome forms the ORFs encoding the replicase, while the remaining one-third encodes the structural and accessory proteins (Fig. 2.2). The ORF1ab at the upstream 5′ vicinity has seven conserved replicase domains whose amino acid sequences, considered as an identity for CoV species, showed 94.4% homology with the same domain of SARS CoV, suggesting the species similarity. (Zhou et al. 2020).

The ORF1ab encodes the large polyprotein complex (PP) which undergoes auto-proteolytic cleavage forming sixteen non-structural proteins (nsps) referred as nsp1 to nsp16. Nsps possess various enzyme activities which are crucial in the life cycle of SARS SoV2 and hence be promising targets for developing antivirals, vaccines and sensors. The structure and the functions of five important nsp enzymes: two viral cysteine proteases, namely nsp3 (papain-like protease) and nsp5 (chymotrypsin-like protease/3CLpro/ main protease), nsp 10 (2′OMTase/Orthomethyl transferase), nsp12 (RNA-dependent RNA polymerase/RdRp) and nsp13 (nucleoside triphosphatase, NTPase /helicase) (Table 2.4) are discussed subsequently as separate subtopics at the end of this chapter, due to their significance as antiviral targets and diagnostic targets.

SARS CoV2 is characterized by an extraordinary mutation rate in their genome, relatively a million times more than their hosts! SARS CoV2 ORF1ab, S, ORF3a, ORF8 and N regions showed many mutations (Wang et al. 2020). Among the five significant non-structural protein genes, the RdRp genes deserve much significance as they are the mutation hotspots which is peculiar in SARS CoV2. RdRP mutation is very remarkable due to the role of RdRP in genome replication and fidelity. Absence of fidelity / or faulty fidelity factors (i.e., faults in proofreading activity and / or faults in post-replicative repair activity) in SARS CoV2 RdRp contributes to the mutagenic competence of SARS CoV2. Mutation in position 14,408 located within the RdRp protein (RdRp interface domain) is considered as a captivating hotspot as it is associated with an overall increased mutation rate. Hence, the RdRp mutation was pinpointed in geographical studies of SARS CoV2 infection. (Pachetti et al. 2020). The source and productivity of the mutated hotspots and the possibility of exploiting them for further research are briefed in Table 2.5.

Table 2.4 Functions of significant SARS CoV2 genes

Gene	Protein/enzyme	Function	References
Nsp1	Functional protein	Blocks the antiviral response of host	Huang et al. (2020)
Nsp2	Functional protein	Function unknown; thought to bind with prohibition protein	Satarker and Nampoothiri (2020)
Nsp3	Papain-like protease	Evasion of host immune system by deubiquitinating enzyme activity and antagonizing interferon response	Dai et al. (2020)
Nsp4	Functional protein	Functions as viroporin	Mittal et al. (2020)
Nsp5	3CLpro main protease	Promoting cytokine expression and cleavage of viral polyprotein	Silva et al. (2020)
Nsp6	Functional protein	Participates in viral replication	Gupta et al. (2020)
Nsp7	Primase	Cofactor for nsp12 Functions as a clamp for RdRp	Huang et al. (2020)
Nsp8	Primase	Forms complex with nsp8 as primase Cofactor for nsp12	Dai et al. (2020)
Nsp 9	Protein phosphatase	RNA-binding protein phosphatase	Davies et al. (2020)
Nsp 10	2′OMTase	Transfers methyl group to viral mRNA and forms the cap	Gupta et al. (2020)
Nsp 11	Peptide	Cofactor for maturation of helicase	Davies et al. (2020)
Nsp12	RdRp	Crucial for viral RNA replication	Silva et al. (2020)
Nsp13	NTPase /helicase	Unwinds viral RNA turns in the presence of Zn ion and assists replication	Huang et al. (2020)
Nsp 14	3′-5 exonuclease	Proofreading of the viral genome post-replication	Huang et al. (2020)
Nsp 15	Uridine-specific Endoribonuclease	Counteracts double-strand RNA sensing Important for sustained replication in the host	Satarker and Nampoothiri (2020)
Nsp 16	RNA capping activity	Cofactor for nsp10 assists in escape from host cell response	Huang et al. (2020)

(continued)

Table 2.4 (continued)

Gene	Protein/enzyme	Function	References
S gene	Spike protein	Viral attachment to host cell receptor	Bosch et al. (2003)
E gene	Envelope protein	Maintaining the shape of virus and viral assembly in the host cytosol	Yan et al. (2020)
M gene	Membrane protein	Organizing the structural proteins	Mittal et al. (2020)
N gene	Nucleocapsid	Signal-dependent packaging of RNA genome within the virions	Huang et al. (2020)

Table 2.5 Types of mutation in SARS CoV2 and their potential uses in COVID-19 management

Mutation type	Consequences of mutation	Opportunities
Error in and/or absence of proofreading activity and post-replicative repair activity	Production of novel mutated variants	To evaluate viral drug resistance
		To elucidate the immune evasion mechanism
Impact of host enzyme	Viral evolution	Repurposing and design of antiviral drugs
Natural physical mutation	Diverse geographical distribution	To formulate vaccine candidates
Natural chemical mutation	Variation in the transmission of the virus	To construct diagnostic sensors
Recombination events	Genome variability. Evasion of host immune response	
	Antiviral drug resistance	

2.2.2 The Structural Protein Genes

Genes encoding the four structural proteins and the accessory putative proteins occur downstream, which unlike ORFab produces the respective proteins via subgenomic mRNA (sgmRNA) intermediates. The common leader sequence of the 5′ end of the parent genome (CLSG) measuring 65–90 nucleotides long is retained in the sgmRNA predicting the discontinuous transcription during template synthesis. CLSG of sgmRNA is called as antileader sequence which is thought to be a promoter sequence for sgmRNA amplification (Pandey et al. 2020; Savarino et al. 2006). Structural proteins (as described in this chapter) are involved in the viral attachment and assembly stage of the life cycle. Putative proteins are primarily involved in host immune evasion, viral virulence and pathogenesis (Bromberg et al. 2007; Cowley et al. 2010). ORF 3b and ORF8 are important accessory proteins (Chan et al. 2020). ORF3b has the capacity to antagonize the INF signaling pathway and inhibit the effector cell activation cascade (the cascade for eradication and inhibition of viral replication, i.e., the cascade for antiviral immune response). ORF6 could bind to the karyopherin-α2 and tether the karyopherin-β1 and consequently inhibit the Janus kinase (JAK)/signal transducer and activator of transcription (STAT) signaling pathways. This leads to suppression of the interferon response (Wang et al. 2017). Both genomic RNAs and sgmRNAs are targets for small interfering RNAs (Wu and Chan 2006) suggesting them as a potential antiviral target.

In addition to the canonical and smRNAs, new transcripts were also identified. These transcripts encode unknown ORF proteins with fusion, deletion and/or frameshift owing to discontinuous transcription events. AAGAA motif and relatively shorter Poy-A tail are modifications sites (Williams et al. 2019; Viehweger et al. 2019; Kim et al. 2020a, b). Deeper understanding of these additional transcripts and modification sites will provide insights into targeting strategies for different stages of SARS CoV2 life cycle.

The SARS CoV2 mRNA has untranslated sequences at the 3′ and 5′ ends, next to the poly-A tail and the leader sequence, respectively. These sequences show around 83% identity with other Betacoronaviruses. They encode the transcriptional regulatory elements which are essential in RNA replication and transcription in coronaviruses. (Savarino et al. 2006). A consensus sequence CUAAAC is found upstream of S protein, M protein and ORF 10 (Wu and Chan 2006). The functions of significant genes of SARS CoV2 genome are summarized in Table 2.4.

2.3 The Non-structural Proteomics

Proteins translated directly from the viral genome (polyprotein complexes) play a promising non-structural role (so-called non-structural proteins or nsps) in the life cycle of coronaviruses. That is, they play pivotal role in the replication and transcription of viral genome to continue the life cycle. Exploring the structure and functions of these proteins may be an important factor in COVID-19 research mainly for the inhibitors-based drug design and vaccine development (Astuti and Ysrafil 2020; Chen et al. 2020). As mentioned in the gene organization, the leading nsps are as follows:

1. PLpro (Papain-like proteases, nsp3).
2. 3CLrpo (Chymotrypsin like proteases, nsp5, also called main proteases).
3. 2OMTase (nsp10).
4. RdRp (RNA-dependent RNA polymerase, nsp12).
5. NTPase/helicase (nsp13).

2.3.1   PLpro

The largest multidomain replicase of SARS CoV is the PLpro. It contains 1922 amino acids, and the active site is formed by a classic catalytic triad of Cys112–His273–Asp287. Functions of PLpro are to process the viral polyprotein in a coordinated manner and deubiquitination. That is, PLpro is involved in stripping ubiquitin and ISG15 (interferon-induced gene 15) from host cell proteins to empower evasion of the host innate immune responses (Báez-Santos et al. 2015; Rut et al. 2020). Owing to its functions, inhibitors of PLpro and nanoparticle conjugates can be used to target SARS CoV2 (Yahira et al. 2015) in order to design new antiviral agents, as discussed in subsequent chapters in this book.

2.3.2   3CLPro—The Main Protease

SARS CoV2 3CLpro is a cysteine protease with the active site conferred by a catalytic diad of Cys 145 and His 41. The high-resolution three-dimensional crystal structure of SARS CoV2 3CLpro is 96% identical to that of the SARS CoV counterpart. It has three structural domains. The first two domains have chymotrypsin-like folding (which is in close identity with α-chymotrypsin, RMSD value-2.6 Å). This folding is well-thought-out as the active catalytic site. The third domain possesses alpha helical motif and is involved in enzyme dimerization rather than catalysis. (Zhang et al. 2020b).

This enzyme is referred as the "main protease" as it catalyzes the hydrolysis and processing of the poly proteins. The 3CLpro is also termed as "the Achilles' heel of coronaviruses" as it is the best therapeutic target in drug repurposing (Yang et al. 2006). Role of 3cLpro in genome replication has insisted researchers to use protease inhibitors such as phytochemicals and nanocarrier-based compounds as an antiviral approach (De Clercq 2006; Ul Qamar et al. 2020), and this has been discussed in subsequent chapters of this book.

2.3.3   2′-OMTase

Silico analysis exposed the conserved fold and the enzyme catalytic tetrad, i.e., Lys46, Asp130, Lys170 and 13 Glu203 of the 2′OMTase (Decroly et al. 2008). 2′OMTase transfers methyl group to the ribose 2′-O position of the first and second nucleotide of viral mRNA cap. This reaction can enable the virus to escape the host immune system recognition. Considering its central role in the viral-host network, 2′OMTase can also be targeted for SARS CoV2 antiviral drug repurposing (Khan et al. 2020).

2.3.4 RdRp

RdRp is the dynamic element of corona viral replication/transcription mechanism. RdRp of SARS CoV2 virus shows 97% sequence homology with that of SARS CoV (Gao et al. 2020). The active site of RdRp is highly conserved beta-turn structure holding two successively protruding aspartate residues which are accessible through the nucleotide channel (Elfiky 2020). Possibility of RdRp active site targeting as strategy for inhibiting viral replication is discussed in Chap. 6. Furthermore, it is appropriate to recall that RdRp is the principal mutation hotspot in SARS CoV2, and the mutants are responsible for the evolution and the host evasion mechanism (Pachetti et al. 2020).

2.3.5 NTPase/Helicase

The nsp13 of SARS CoV2 exhibit dual activities of both the nucleoside triphosphate hydrolase (NTPase) and RNA helicase capable of hydrolyzing all types of NTPs and unwinding RNA helical turns dependently in the presence of NTP. High homology of the amino acid sequences among corona viral nsp13 proteins is worth mentioning here (Shu et al. 2020). The active site of SARS CoV2 NTPase/helicase contains Ser310, Lys288 and Glu375 (involved in H-bonding), Arg178 and conserved alanine residues (Ala312 and Ala314) which contribute significantly toward the total binding free energy (45 Jia et al. 2019). NTPase/helicase of SARS CoV is hypothesized to execute crucial roles in the viral life cycle, making it a striking target for anti-COVID-19 therapy. (Lee et al. 2009). Therefore, drugs repurposed for inhibiting the NTPase/helicase activity of nsp13 can be potent against COVID-19 for timely treatment. (Mirza and Froeyen 2020).

Until now, we have discussed on significant viral proteomics; hitherto, the host cell proteomics is also equally important to understand the cell's antiviral and immune mechanism against infection. Translatome and proteome proteomics scrutiny in SARS CoV2-infected human host cell culture illustrates that translation, splicing, carbon metabolism and nucleic acid metabolism are host target pathways attacked by the virus. These pathways can be potentially be inhibited by small molecule inhibitors, for getting insights into drugs and vaccines candidates to fight COVID-19 (Bojkova et al. 2020). Expression of SARS CoV2 proteins in human cells as analyzed by the affinity-purification mass spectrometry let drops SARS CoV2 vs human protein–protein interactions and predicts 66 druggable human proteins or host factors (Gordon et al. 2020).

References

Andersen KG, Rambaut A, Lipkin WI, Holmes EC, Garry RF. The proximal origin of SARS-CoV-2. Nat Med. 2020;26:450–2.

Astuti I, Ysrafil. Severe acute respiratory syndrome coronavirus 2 (SARS-CoV-2): an overview of viral structure and host response. Diab Metab Syndr. 2020;14(4):407–412.

Báez-Santos YM, St John SE, Mesecar AD. The SARS-coronavirus papain-like protease: structure, function and inhibition by designed antiviral compounds. Antiviral Res. 2015;115:21–38.

Bianchi M, Benvenuto D, Giovanetti M, Angeletti S, Ciccozzi M, Pascarella S. Sars-CoV-2 envelope and membrane proteins: structural differences linked to virus characteristics? BioMed Res Int. 2020:6. Article ID 4389089.

Bojkova D, Klann K, Koch B, Widera M, Krause D, Ciesek S, Cinatl J, Münch C. Proteomics of SARS-CoV-2-infected host cells reveals therapy targets. Nature. 2020;583:469–72.

Bosch BJ, van der Zee R, de Haan CA, Rottier PJ. The coronavirus spike protein is a class I virus fusion protein: structural and functional characterization of the fusion core complex. J Virol. 2003;77:8801–11.

Bromberg SA, Martinez-Sobrido L, Frieman M, Baric RA, Palese P. Severe acute respiratory syndrome coronavirus open reading frame (ORF) 3b, ORF 6, and nucleocapsid proteins function as interferon antagonists. J Virol. 2007;81:548–57.

Chan JF, Yuan S, Kok KH, Wang To KK, Chu H, Yang J, Xing F, Liu J, Yan Yip CC, Shan Poon RW, Tsoi HW, Fai Lo SK, Chan KH, Poon VK, Chan WM, Daniel J, Cai JP, Cheng VCC, Chen H, Hui CKM. A familial cluster of pneumonia associated with the 2019 novel coronavirus indicating person-to-person transmission: a study of a family cluster. Lancet. 2020;395(10233):514–23.

Chen Y, Liu Q, Guo D. Emerging coronaviruses: genome structure, replication, and pathogenesis. Joournal of Medical Virology. 2020;92(4):418–23.

Cowley TJ, Long SY, Weiss SR. The murine coronavirus nucleocapsid gene is a determinant of virulence. J Virol. 2010;84:1752–63.

Dai W, Zhang B, Su H, et al. Structure-based design of antiviral drug candidates targeting the SARS-CoV-2 main protease. Science. 2020;80:1335.

Davies JP, Almasy KM, McDonald EF, Plate L. Comparative multiplexed interactomics of SARS-CoV-2 and homologous coronavirus non-structural proteins identifies unique and shared host-cell dependencies. Preprint. bioRxiv. 2020;2020.07.13.201517. Published 2020 Jul 14. doi:https://doi.org/10.1101/2020.07.13.201517

De Clercq E: Potential antivirals and antiviral strategies against SARS coronavirus infections. Expert Rev Anti Infect Ther. 2006;4(2):291–302.

Decroly E, Imbert I, Coutard B, Bouvet M, Selisko B, Alvarez K, Gorbalenya AE, Snijder EJ, Canard B. Coronavirus non-structural protein 16 is a cap-0 binding enzyme possessing (nucleoside-2′O)-methyltransferase activity. J Virol. 2008;82(16):8071–84.

Elfiky AA. Ribavirin, Remdesivir, Sofosbuvir, Galidesivir, and Tenofovir against SARS-CoV-2 RNA dependent RNA polymerase (RdRp): A molecular docking study. Life Sci. 2020;253:117592.

Gao Y, Yan L, Huang Y, Liu F, Zhao Y, Cao L, Wang T, Sun Q, Ming Z, Zhang L, Ge J, Zheng L, Zhang Y, Wang H, Zhu Y, Zhu C, Hu T, Hua T, Zhang B, Rao Z. Structure of the RNA-dependent RNA polymerase from COVID-19 virus. Science. 2020;368(6492):779–82.

Gordon DE, Jang GM, Bouhaddou MA, et al. SARS-CoV-2 protein interaction map reveals targets for drug repurposing. Nature. 2020;583:459–68.

Gupta R, Paswan RR, SAikia R, Borar BK. Insights into the severe acute respiratory syndrome coronavirus-2: transmission, genome composition, replication, diagnostics and therapeutics. Curr J Appl Sci Technol. 2020;39(21):71–91.

Huang Y, Yang C, Xu X, et al. Structural and functional properties of SARS-CoV-2 spike protein: potential antivirus drug development for COVID-19. Acta Pharmacologica Sinica B. 2020;41:1141–9.

Jia Z, Yan L, Ren Z, Wu L, Wang J, Guo J, Zheng L, Ming Z, Zhang L, Lou Z, Rao Z. Delicate structural coordination of the severe acute respiratory syndrome coronavirus Nsp13 upon ATP hydrolysis. Nucleic Acids Res. 2019;47(12):6538–50.

Jiang S, Hillyer C, Du L. Neutralizing antibodies against SARS-CoV-2 and other human Coronaviruses. Trends Immunol. 2020;41(5):355–9.

Jimenez-Guardeno JM, Nieto-Torres JL, DeDiego ML, Regla-Nava JA, Fernandez-Delgado R, Castano-Rodriguez C, Enjuanes L. The PDZ-binding motif of severe acute respiratory syndrome coronavirus envelope protein is a determinant of viral pathogenesis. PLoS Pathog. 2014;10(8):e1004320.

Khan RJ., Jha RK, Amera GM, Jain M, Singh E, Pathak A., Singh RP, Muthukumaran J, Singh A K. Targeting SARS-CoV-2: a systematic drug repurposing approach to identify promising inhibitors against 3C-like proteinase and 2'-O-ribose methyltransferase. Journal of biomolecular structure & dynamics 2020; 1–14. Advance online publication. https://doi.org/10.1080/07391102.2020. 1753577

Kim J, Lee Y, Yang JSS, Kim JW, Kim VN, Chang H. The architecture of SARS-CoV-2 transcriptome. Cell. 2020;181(4):914–21.

Kim JM, Chung YS, Jo HJ, Lee NJ, Kim MS, Woo SH, Park S, Kim JW, Kim HM, Han MG. Identification of coronavirus isolated from a patient in Korea with COVID-19. Osong Publ Health Res Persp. 2020;11:3–7.

Lee C, Lee JM, Lee N, Kim DE, Chong Y. Investigation of the pharmacophore space of severe acute respiratory syndrome coronavirus (SARS-CoV) NTPase/helicase by dihydroxychromone derivatives. Bioorg Med Chem Lett. 2009;19(1615):4538–4541.

Mirza, MU, Froeyen M. Structural elucidation of SARS-CoV-2 vital proteins: Computational methods reveal potential drug candidates against main protease. Nsp12 RNA-dependent RNA polymerase and Nsp13 helicase. J Pharm Anal. 2020; In press.

Mittal A, Manjunath K, Ranjan RK, Kaushik S, Kumar S, Verma V. COVID-19 pandemic: Insights into structure, function, and hACE2 receptor recognition by SARS-CoV-2. PLoS Pathog. 2020;16(8):e1008762.

Narayanan K, Makino S. Cooperation of an RNA packaging signal and a viral envelope protein in coronavirus RNA packaging. J Virol. 2001;75(19):9059–67.

Pachetti M, Marini B, Benedetti F, Giudici F, Mauro E, Storici P, Masciovecchio C, Angeletti S, Ciccozzi M, Gallo RC, Zella D, Ippodrino R. Emerging SARS-CoV-2 mutation hot spots include a novel RNA-dependent-RNA polymerase variant. J Trans Med. 2020;18(1):179.

Pandey A, Nikam AN, Shreya AB, Mutalik SP, Gopalan D, Kulkarni S, Padya BS, Fernandes G, Mutalik S, Prassl R. Potential therapeutic targets for combating SARS-CoV-2: drug repurposing, clinical trials and recent advancements. Life Sci. 2020;256:117883.

Rut W, Żmudziński M, Snipas SJ, Bekes M, Huang TT, Drag M. Engineered unnatural ubiquitin for optimal detection of deubiquitinating enzymes. Chem Sci. 2020;11:6058–69.

Satarker S, Nampoothiri M. Structural proteins in severe acute respiratory syndrome coronavirus-2. Archives of Medicaa Research. 2020.

Savarino A, Buonavoglia C, Norelli S, Trani LD, Cassone A. Potential therapies for coronaviruses. Expert Opin Ther Pat. 2006;16(9):1269–88.

Schoeman D, Fielding BC. Coronavirus envelope protein: current knowledge. Virology J. 2019;16(1):P-NA.

Shereen MA, Khan S, Kazmi A, Basheer N, Siddique R. COVID-19 infection: origin, transmission, and characteristics of human coronaviruses. J Adv Res. 2020;24:91–8.

Shu T., Huang M, Wu D, Ren Y, Zhang X, Han Y, Mu J, Wang R, Qiu Y, Zhang DY, Zhou X. SARS-coronavirus-2 Nsp13 possesses NTPase and RNA helicase activities that can be inhibited by Bismuth salts. Virologica Sinica. 2020;1–9. Advance online publication: https://doi.org/https://doi.org/10.1007/s12250-020-00242-1.

Silva SJR, Alves da Silva CT, Mendes RPG, Pena L. Role of nonstructural proteins in the pathogenesis of SARS-CoV-2. J Med Virol 2020;92:1427–1429.

Tai W, He L, Zhang X, Pu J, Voronin D, Jiang S. Characterization of the receptor-binding domain (RBD) of 2019 novel coronavirus: implication for development of RBD protein as a viral attachment inhibitor and vaccine. Cell Mol Immunol. 2020;17:613–20.

Ul Qamar M.T, Alqahtani SM, Alamri MA, Chen LL. Structural basis of SARS-CoV-2 3CLpro and anti-COVID-19 drug discovery from medicinal plants. J Pharm Anal. 2020. doi:https://doi.org/10.1016/j.jpha.2020.03.009.

Viehweger A, Krautwurst S, Lamkiewicz K, Madhugiri R, Ziebuhr J, Hölzer M, Marz M. Direct RNA nanopore sequencing of full-length coronavirus genomes provides novel insights into structural variants and enables modification analysis. Genome Res. 2019;29:1545–54.

Walls AC, Park YJ, Tortorici MA, Wall A, McGuire AT, Veesler D. Structure, function, and antigenicity of the SARS-CoV-2 spike glycoprotein. Cell. 2020;181(2):281-292.e6.

Wang C, Liu Z, Chen Z, Huang X, Xu M, He T, Zhang Z. The establishment of reference sequence for SARS-CoV-2 and variation analysis. J Med Virol. 2020;92(6):667–74.

Wang C, Sun M, Yuan X, et al. Enterovirus 71 suppresses interferon responses by blocking Janus kinase (JAK)/signal transducer and activator of transcription (STAT) signaling through inducing karyopherin-α1 degradation. J Biol Chem. 2017;292(24):10262–74.

Williams GD, Gokhale NS, Horner SM. Regulation of viral infection by the RNA modification N6-methyladenosine. Ann Rev Virol. 2019;6:235–53.

Wu CJ, Chan YL. Antiviral applications of RNAi for coronavirus. Expert Opin Investig Drugs. 2006;15(2):89–97.

Wu F, Zhao S, Yu B, Chen YM, Wang W, Song ZG. A new coronavirus associated with human respiratory disease in China. Nature. 2020;579:265–9.

Yahira M, Santos h B, Andrew DM. The SARS-coronavirus papain-like protease: structure, function and inhibition by designed antiviral compounds. Antiviral Res. 2015;115:21–38.

Yan R, Zhang Y, Li Y, Xia L, Guo Y, Zhou Q. Structural basis for the recognition of SARS-CoV-2 by full-length human ACE2. Science. 2020;367:1444–8.

Yang H, Bartlam M, Rao Z. Drug design targeting the main protease, the achilles heel of coronaviruses. Curr Pharm Des. 2006;12:4573–90.

Zhang J, Zeng H, Gu J, Li H, Zheng L, Zou Q. Progress and prospects on vaccine development against SARS-CoV-2. Vaccines. 2020;8(2):153.

Zhang L, Lin D, Sun X, Curth, Drosten C, Sauerhering L, Becker S, Rox K, Hilgenfeld R. Crystal structure of SARS-CoV-2 main protease provides a basis for design of improved α-ketoamide inhibitors. Science. 2020;368(6489):409–412.

Zhou P, Yang XL, Wang XG, Hu B, Zhang L, Zhang W, Si HR, Zhu Y, Li B, Huang CL, Chen HD, Chen J, Luo Y, Guo H, Jiang RD, Liu MQ, Chen Y, Shen XR, Wang X, Zheng XS, Zhao K, Chen QJ, Deng F, Liu LL, Yan B, Zhan FX, Wang YY, Xiao GF, Shi ZL. A pneumonia outbreak associated with a new coronavirus of probable bat origin. Nature. 2020;579:270–3.

Chapter 3
Tropism of SARS CoV2

With the knowledge on the structure and functions of SARS CoV2 and the host protein receptor (as discussed in Chap. 2), it is easy to understand the stages in the life cycle of coronaviruses. This is because proteins, genes and receptors are chief components in every stage of the life cycle. Also, the analysis of stages of life cycle reveals possible inhibiting targets (Brockway and Denison 2004; Lin et al. 2017; Zhavoronkov et al. 2020), crucial for different modalities of COVID-19 management such as therapeutics, vaccine development and sensor fabrication (detailed in Chap. 4).

3.1 Eight Stages of SARS CoV2 Life Cycle

The eight stages of SARS CoV2 life cycle (Savarino et al. 2006; Shereen et al. 2020) are described in this chapter and illustrated in Fig. 3.1 and Table 3.1.

3.1.1 STAGE 1: Viral Attachment to Host Cell Membrane and Entry into Cytoplasm

SARS CoV2 gains entry into the host cytoplasm by one of the two mechanisms: (i) the endosomal endocytosis and (ii) the plasma membrane fusion, both mediated by the spike glycoprotein (Wrapp et al. 2020).

The first mechanism depends on ACE2 receptor. The S1 domain of the spike glycoprotein binds to the ACE2 receptor of the host cell to form virus-receptor complex. This complex is surrounded by membrane to form endosome which is internalized into the cytoplasm by endocytosis. Endocytosis is favored by the acidic pH. In the cytoplasm, the virus envelope fuses with the endosomal membrane, releasing the positive-sense RNA and the nucleocapsid which is mediated by the S2 domain.

Devasena T., *Nanotechnology-COVID-19 Interface*, Nanotheranostics, https://doi.org/10.1007/978-981-33-6300-7_3

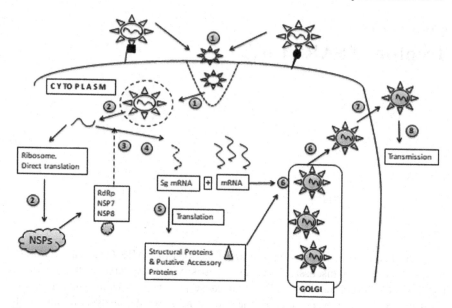

Fig. 3.1 Life cycle of SARS CoV2. Legend: (1) Viral attachment and endosome formation. (2) Direct translation and formation of non-structural proteins. (3) Replication of viral RNA. (4) Formation of m RNA and subgenomic mRNA. (5) Formation of structural and accessory proteins. (6) Assembly of proteins and mRNAs into new virions. (7) Budding and gemmation of new virions out of host cell. (8) Transmission to continue the life cycle

Table 3.1 Stages of SARS CoV2 life cycle

Stages of life cycle	Location in host cell	Factors involved
Viral attachment and entry	Membrane	Spike protein of virus Furin protease, TMPRSS and Cathepsin of the host cell
Direct translation of the 5′ end ORF of the internalized viral genome	Cytosol, ribosome	Ribosomes and endoplasmic reticulum machinery
Polymerization/replication of viral RNA (the mRNA)	Nucleus	Nsp12, nsp 7 and nsp8
Discontinuous transcription of viral genome into subgenome-sized RNA (the sgmRNA)	Nucleus	Nsp12 RdRP activity
Translation of sgmRNAs into structural and accessory proteins	Cytosol, ribosome	
Organization assembly and packaging of proteins and new RNAs	Golgi apparatus	M protein and N protein of the virus
Budding and release of new viral particles	Plasma membrane	Endosomes
Transmission	Extracellular	Respiratory and speech aerosols

Asian statistics reveal that SARS CoV2 infection exhibits male preponderance due to higher expression of ACE2 receptor, suggesting the crucial role of ACE2 in the viral entry. The receptor-binding domain of the spike protein of the SARS CoV2 has higher affinity for ACE2 as compared to SARS CoV which confirms higher contagion of the former (Lu et al. 2020; Zheng 2020; Wang et al. 2020).

In the second mechanism, the virus depends additionally on the host enzyme factors: the cathepsin and TMPRSS2 (the transmembrane serine protease 2; Shirato et al. 2018) and furin protease (Hoffmann et al. 2020; Xia et al. 2020). These proteins catalyze the cleavage of S protein at the multibasic S1-S2 interface and expose the fusion peptide thereby enabling the direct fusion of virus and the host cell membrane. After fusion, the genome is released down into the cytoplasm. Priming by TMPRSS2 is responsible for increasing the viral spread (Liu et al. 2010).

Recently, spike proteins were reported to gain host cell entry by an ACE2-independent lectin pathway. This shows that SARS CoV2 spike protein has glycan-heterogeneity which enables binding to host C-type lectins and siglecs (sialic acid-binding immunoglobulin-type lectins) (Chiodo et al. 2020).

3.1.2 Stage 2: Direct Translation of Viral ORF1ab Gene in the Host Cytoplasm

The ORF 1ab (Refer Chap. 2, Fig. 2.2) is directly translated in the host cytoplasm to form polyproteins or replicase complex. The replicase complex undergoes auto-proteolytic cleavage by viral-encoded protein containing PLro and 3CLpro activity. This cleavage process generates non-structural proteins (nsp1 to nsp16) which forms the replicase-transcriptase complex (RTC) (Astuti et al. 2020).

3.1.3 Stage 3: Replication/Polymerization of Viral RNA by Nsps

Replication in SARS CoV2 is the process whereby genome-sized RNA (which also functions as mRNA) is produced. The nsp12 (the main polymerizing enzyme with RdRP activity) along with nsp7 and nsp8 (as cofactors) replicates the whole viral genome and produces multiple copies which will be ready for packaging into virions (Fehr and Perlman 2015; Romanao et al. 2020).

3.1.4   Stage 4: Formation of Subgenomic mRNAs by Discontinuous Transcription of RNA Genome

In SARS CoV2 life cycle, only certain sequences of the canonical genome are transcribed into sgmRNAs; and hence the process is called discontinuous transcription. It is valid to note that during replication (stage 3) the genome-sized RNAs are produced while during incomplete transcription (stage 4), subgenome-sized RNAs are produced. RdRP is involved in this process.

3.1.5   Stage 5: Translation of SgmRNAs into Structural and Accessory Proteins

SgmRNAs are translated to structural proteins (S, E, M and N proteins) and putative accessory proteins (Kim et al. 2020).

3.1.6   Stage 6: Organization and Assembly of Proteins and New RNAs

RNA copies formed in stage 3 and proteins formed in stage 5 are organized and assembled into complete virion in the golgi and released back into the cytoplasm (Kumar et al. 2020).

3.1.7   Stage 7: Budding and Release of New Viral Particles

New virions in the cytoplasm get encapsulated in endosomes and transported to the host cell membrane where they bud off and get released out by exocytosis. After the release, the virus gets transmitted to invade other host cells (He et al. 2020).

3.1.8   Stage 8: Transmission

Released SARS CoV2 particles can survive in the air for 2 h during which the primary transmission is mediated by respiratory droplets and speech droplets to infect other hosts by, cough, sneezing or speech. Touching of infected surface and transmission to the eyes, nose or mouth are also other sources for infection. Conjunctival epithelium is easily infected by SARS CoV2. The incubation time of SARS CoV2 after is 4 to 8 days post-infection (Feng et al. 2020; Zhang 2020; Stadnytskyia et al. 2020).

References

Astuti IY. Severe acute respiratory syndrome coronavirus 2 (SARS-CoV-2): an overview of viral structure and host response. Diab Metab Syndr. 2020;14(4):407–412.

Brockway SM, Denison MR. Molecular targets for the rational design of drugs to inhibit SARS coronavirus. Drug Discov Today Dis Mech. 2004;1(2):205–9.

Chiodo F, Bruijns SCM, Rodriguez E, Eveline Li RJ, Molinaro A, Silipo A, Di Lorenzo F, Garcia-Rivera D, Valdes-Balbin Y, VerezBencomo V, Kooyk YV. Novel ACE2-Independent Carbohydrate-Binding of SARS-CoV-2 Spike Protein to Host Lectins and Lung Microbiota. bioRxiv preprint 2020; doi: https://doi.org/https://doi.org/10.1101/2020.05.13.092478.

Fehr AR, Perlman S. Chapter 1. Coronaviruses: an overview of their replication and pathogenesis. In: Maier HJ et al., editors. Coronaviruses: Methods and Protocols, Methods in Molecular Biology. 2015; 1282: 1–23. DOI https://doi.org/10.1007/978-1-4939-2438-7_1. Springer, New York.

Feng W, Zong W, Wang F. Ju S. Severe acute respiratory syndrome coronavirus 2 (SARS-CoV-2): a review. Molecular Cancer. 2020;19:100.

He F, Deng Y, Li W. Coronavirus disease 2019: what we know? J Med Virol. 2020;92(7):719–25.

Hoffmann M, Kleine-Weber H, Schroeder S, Krüger N, Herrler T, Herichsen S, Schiergens, Herrler G, Wu NH, Nitsche, Marcel A Müller MA, Drosten C, Pöhlmann S. SARS-CoV-2 cell entry depends on ACE2 and TMPRSS2 and is blocked by a clinically proven protease inhibitor. Cell. 2020;181:271–280.

Kim D, Lee JY, Yang JS, Kim JW, Kim VN, Chang H. The architecture of SARS-CoV-2 transcriptome. Cell. 2020;181 (4):914–921.

Kumar S, Nyodu R, Maurya VK, Saxena SK. Morphology, genome organization, replication, and pathogenesis of severe acute respiratory syndrome coronavirus 2 (SARS-CoV-2). In: Saxena S, editors. Coronavirus Disease 2019 (COVID-19). Medical Virology: From Pathogenesis to Disease Control. 2020; Springer, Singapore.

Lin SC, Ho CT, Chuo WH, Li S, Wang TT, Lin CC. Effective inhibition of MERS-CoV infection by resveratrol. BMC Infect Dis. 2017;17(1):144.

Liu T, Luo S, Libby P, Shi GP. Cathepsin L-selective inhibitors: a potentially promising treatment for COVID-19 patients. Pharmacol Ther. 2010;126(3):251–62.

Lu R, Zhao X, Li J, Peihua N, Bo Y, Honglong W. Genomic characterisation and epidemiology of 2019 novel coronavirus: implications for virus origins and receptor binding. Lancet. 2020;395:565–74.

Romanao M, Ruggiero A, Squeglia F, Maga G, Berisio R. A structural view of SARS-CoV-2 RNA replication machinery: RNA synthesis, proofreading and final capping. Cells. 2020;9(5):1267.

Savarino A, Buonavoglia C, Norelli S, Trani LD, Cassone A. Potential therapies for coronaviruses. Expert Opin Ther Pat. 2006;16(9):1269–88.

Shereen MA, Khan S, Kazmi A, Basheer N, Siddique R. COVID-19 infection: Origin, transmission, and characteristics of human coronaviruses. J Adv Res. 2020;24:91–8.

Shirato K, Kawase M, Matsuyama S. Wild-type human coronaviruses prefer cell-surface TMPRSS2 to endosomal cathepsins for cell entry. Virology. 2018;517:9–15.

Stadnytskyia V, Baxb CE, Baxa A, Anfinruda P. The airborne lifetime of small speech droplets and their potential importance in SARS-CoV-2 transmission. 2020;367(6483):1260–1263.

Wang D, Hu B, Hu C, et al. Clinical characteristics of 138 hospitalized patients with 2019 novel coronavirus–infected pneumonia in Wuhan China. JAMA. 2020;323(11):1061–9.

Xia, S., Lan, Q., Su, S. Wang X, Liu Z, Zhu Y,Wang Q, Lu L, Jiang S. The role of furin cleavage site in SARS-CoV-2 spike protein-mediated membrane fusion in the presence or absence of trypsin. Signal Transduction Targeted Ther. 2020;5, Article No: 92.

Zhang DX. SARS-CoV-2: air/aerosols and surfaces in laboratory and clinical settings. J Hosp Infect. 2020;105(3):577–579.

Zhavoronkov A, Aladinskiy V, Zhebrak A, Zagribelnyy B. Terentiev V, Bezrukov DS. Potential COVID-2019 3C-like protease inhibitors designed using generative deep learning approaches. ChemRxiv. 2020; doi: https://doi.org/10.26434/chemrxiv.11829102.V1.

Zheng J. SARS-CoV-2: an emerging coronavirus that causes a global threat. Int J Biol Sci. 2020;16(10):1678.

Chapter 4
The Nanotechnology-COVID-19 Interface

4.1 Nanotechnology for COVID-19 Management

The SARS CoV2 pandemic has raised the following global demands:

(i)   Effective drugs to target the SARS CoV2 life cycle of the host receptor
(ii) Specific vaccine candidate for prevention
(iii) Sensitive and speedy diagnostic tool
(iv) High-quality personal protective equipments (PPEs).

Though already marketed antiviral drugs are used in the research to treat SARS CoV2 infection, the FDA approved drug and vaccine is still under different phases of clinical trials (Tse et al. 2020). As of now, convalescent plasma therapy (Abraham 2020) nanobodies (Wrappet al. 2020; McKee et al. 2020) and the drug chloroquine are the frontliners in the therapeutic option for SARS CoV2 infection while the other choices are SiRNA (Ghosh et al. 2020), viral attachment and entry inhibitors (Verdecchia et al. 2020; Shetty et al. 2020) viral genome replication inhibitors(Caly et al. 2020), transcription inhibitors (Frieman et al. 2007; Zumla et al. 2016), nucleoside analogs, inhibitors of inflammatory response and cytokine storm and Chinese traditional medicine (Lu 2020). Similarly, when the detection mode is considered, only PCR assays and CT scanning are available, and the method for early, specific and sensitive detection of COVID-19 is still debatable. The prevention mode by SARS CoV2 vaccines is also in the stage of infancy. The need for high-quality PPEs capable of inactivating and filtering the viruses from entering into the body is also constantly increasing.

In this critical pandemic situation, it is valuable to exploit the preeminence and innovations in nanotechnology for deciphering the SARS CoV2-nanotechnology interface and circumvent the existing demerits in facing the pandemic and finally achieve the above-mentioned global demands (Uskokovic 2007; Nasrollahzadeh et al. 2020). Various nanotechnology concepts and nanomaterials can be translated to improvise the already existing strategies as well as in developing novel innovative strategies. This will enable faster and sensitive diagnosis, effective treatment

Devasena T., *Nanotechnology-COVID-19 Interface*, Nanotheranostics, https://doi.org/10.1007/978-981-33-6300-7_4

and timely prevention of COVID-19 (Singh et al. 2017; Chan 2020). Indisputably, nanomedicine is a novel and precise avenue for the discovery of value-added candidates for the diagnosis, treatment and prevention of COVID-19. For example, multi-functionalized delivery systems such as metal nanoparticles, polymeric nanocapsules, lipid vesicles, dendrimers, micelles and inorganic nanomaterials can be surface engineered to serve the purpose of efficient anti-SARS CoV2 therapy (Lembo and Cavalli 2010; Boles et al. 2016). Surface-engineered nanosystems possess the advantages of better release kinetics, higher bioavailability, targeted delivery of drugs, better immune response by vaccines, low toxicity and better prognosis (Nikaeen et al. 2020; Udugama et al. 2020). From the prevention point of view, nanoparticles and nanofibers can play role in textile coating in order to fabricate high-quality fabrics for personal protective equipments (Sportelli et al. 2020). Many challenges however must be addressed prior to the translation of nanotechnology into safe and effective anti-SARS CoV2 formulations for clinical use (Sivasankarapillai et al. 2020).

4.2 Nanoinsights into the Four Modalities of COVID-19 Management

Deciphering the nanotechnology–virology interface can to a certain extent link the gap between the vital phases of SARS CoV2 life cycle and the four modalities of COVID-19 management (sensing/diagnosis, therapy, prevention and self-protection).

(a) **The sensing modality**

SARS CoV2 can be sensed in a given sample by analyzing the presence of specific target proteins (the SARS CoV2 antigens, the virulent epitopes, the anti-SARS CoV2 antibodies) or target oligonucleotides (specific sequences of canonical or sgmRNAs). But, immaterial of the targets analyzed, the detection method should provide high specificity, sensitivity and signal amplification. All these criteria can be fulfilled by the advancements in nanotechnology that could be used in diagnostic tools and sensing materials. In fact, some nanomaterials have already been proved to be useful in the detection of SARS CoV and MERS CoV. Among the sensing materials, graphene deserves much significance because of two reasons: (i) the extraordinary surface area and electrical properties that contribute to high sensitivity and (ii) surface engineering of GO with proteins, enzymes, oligonucleotides and small molecules is feasible due to high density of surface oxygen. GO is capable of detecting the antigenic determinants of SARS CoV2 (Chauhan et al. 2020; Palmieri and Papi 2020). Secondly, the quantum dots are good choice for the detection of SARS CoV antigens (Roh and Jo 2011; Ahmed et al. 2018). Third, metal nanoparticles like superparamagnetic iron oxide nanoparticles (SPIONS), silver and gold can be used as detection material for SARS CoV2 because they have potential to sense oligonucleotide target, E protein and N protein targets of SARS CoV and MERS CoV (Gong et al. 2008; Kim et al. 2019; Teengam et al. 2017).

(b) **The Therapeutic Modality**

Nanomaterials have already been used as a therapeutic strategy to combat coronavirus and other enveloped positive-sense RNA viruses which are similar to coronavirus. The same concept can be exploited for the design and development of anti-SARS CoV2 drugs which can hit the following targets:

- The viral spike protein structure and attachment to the host
- The host cell receptors involved in the viral entry
- The acid proteases that enable viral release into the host cytoplasm
- The canonical genome replication
- The sgmRNA synthesis.
  The unique properties of nanomaterial such as (i) possibility for surface engineering (ii) targeted and sustained drug release (iii) high bioavailability of the cargo and iv) better site-specific efficacy will contribute to better therapy (Chauhan et al. 2020).

(c) **The preventive modality**

Viral vaccines constructed out of antigenic epitopes, specific amino acid sequences, fusion proteins and RNA oligonucleotides come under preventive modality. Though the protein vaccines, peptide vaccines and RNA vaccines are capable of eliciting immune response by raising antibodies in the host cell, the cytokine storm and side effects should be considered as an antisafety factor (Chauhan et al. 2020). In this context, nanoparticles can be used as best adjuvants for viral vaccines which will not only reduce the cytokine storm and side effects but also enhance the antibody titer. Hence, nanotechnology-based vaccines prepared from spike proteins or other virulent proteins may be a promising and safe strategy to prevent SARS CoV2 infection (Yue et al. 2012). Already, micellular spike protein nanovaccines have been developed for SARS CoV and MERS CoV. This vaccine gives a high antibody response in conjunction with saponin-based adjuvant (Coleman et al. 2017). Spike protein-bound gold nanoparticles elicit T cell immune response with high efficacy against coronavirus infection (Chen et al. 2016a, b). Secondly, the SARS CoV N protein vaccine can be coupled with chitosan nanoparticles with positive surface charge to target the lung epithelial cell membrane with negative surface charge. (Raghuwanshi et al. 2012). As S protein and N protein are crucial for viral attachment and RNA packaging in SARS CoV2, nanoparticles may be engaged to develop S protein and N protein vaccines for preventing COVID-19 disease.

(d) **The protective modality**

PPEs like face mask, hand hygiene products, face shield, goggles, gloves, gowns and shoe covers come under protective modality. Face mask and hand hygiene products are however very important for the common people while others are essential for the frontline workers in environmental and healthcare settings. Respiratory and speech aerosols containing the viral particles in transmission mode can be captured and inactivated by nanoparticles incorporated into the textile material of the PPEs.

Face mask textile materials coated with graphene, and copper oxide nanoparticles can also be protective (Li et al. 2020; Borkow et al. 2010). Silica composites exhibit virucidal effect when deposited on facial masks designed specifically for SARS CoV2. This type of coating has the ability to provide protection when coated to filtering media and also on metallic, ceramic, polymeric and glass surfaces. This coating is also free of undesirable waste generation during disposal, and it is highly safe to be used in crowded places where surfaces are exposed to many contacts (Balagna et al. 2020). Silver nanoparticles, silicon nanostructures and titania functionalized electrospun nanofibers have the ability to retain the atmospheric viral aerosols and destroy them in the presence of radiation (Joe et al. 2016; El-Atab et al. 2020; Lee et al. 2010; Konda 2020).

Therefore, value-added disinfective PPEs can be fabricated using nanotechnology solutions in order to prevent the entry of SARS CoV2. There are evidences for the underuse of nanotechnology in SARS CoV2 research, inadequate number of publications and insufficient patents on coronavirus–nanotechnology interface (Uskokovic 2007; Kostarelos et al. 2020). Hence, the succeeding topics throw a distinctive emphasis on the existing and promising role of potential theragnostic nanomaterials which may be of immense use in the management strategies for SARS CoV2 pandemic. This interfacial approach can lead to the prompt design of nanomaterial-based therapeutics, vaccines, sensors and PPEs of high values.

4.3   Potential Theranostic Nanomaterials for Managing COVID-19

Following are the unique properties of nanoparticles that contribute to their antiviral properties:

- Small size which leads to deeper penetration in the target site for diagnosis and imaging
- High surface to volume ratio that permits functionalization of ligands of medical value such as drugs, imaging agents, targeting agents, immune evasion agents, stabilizing agents and allows encapsulation of drugs with high loading efficiency and controlled release property
- Modifiable surface which enables surface engineering resulting in multifunctional nanosystem which can load cargos like drugs, targeting groups, imaging agents, stabilizing ligands, etc.
- Controllable solubility (hydrophobicity or hydrophilicity)
- Tunable surface charge and size that enable drug delivery into precise nanoscale target sites
- Enhanced circulatory half-life

Table 4.1 Types of nanomaterials with potential applications in the management of SARS CoV2 infection

Nanomaterials	Examples	References
Organic nanocarriers	Dendrimers Liposomes Nanomicelles Solid-lipid nanoparticles	Itani et al. (2020)
Polymeric nanoparticles	Chitosan nanocapsules PEG-PLGA nanoparticles	Itani et al. (2020)
Carbon nanomaterials	Nanodiamond Graphene oxide nanosheets Fullerene Carbon dots	Bhavana et al. (2020)
Inorganic oxide nanoparticles	Zirconia nanoparticles Titanium oxide nanoparticles Zinc oxide nanoparticles	Itani et al. (2020)
Metal nanoparticles	Gold nanoparticles Silver nanoparticles Copper nanoparticles Platinum nanoparticles	Lv et al. (2014); Ruiz-Hitzky et al. (2020)
Quantum dots	Cadmium telluride Cadmium selenide Cadmium sulfide Zinc sulfide Zinc selenide Curcumin-based quantum dots Silicon	Bhavana et al. (2020); Singh et al. (2017); Ting et al. (2018a, b); Lin et al. (2019); Sing Ashish et al. (2017)

- Tunable optical properties which enable high-resolution contrast imaging in diagnosis.

Various nanomaterials with promising applications in COVID-19 nanomedicine are shown in Table 4.1.

4.3.1 Organic Nanocarriers for Antiviral Drug/vaccine Delivery

Dendrimers, liposomes, nanomicelles, solid-lipid nanoparticles (SLNs) (Fig. 4.1.)

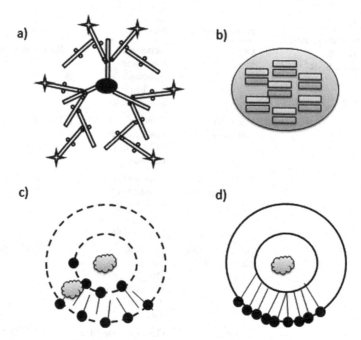

Fig. 4.1 Nanocarriers for anti-SARS CoV2 drugs: **a** Dendrimer **b** solid-lipid nanoparticle **c** liposomes **d** Nanomicelles

and chitosan nanocapsules are lead organic nanomaterials used in the controlled release of antiviral drugs.

Dendrimers

Dendrimers are hyperbranched polymers with core, dendrites or branches (called generations, G) and highly tunable polyvalent surfaces. The diameter of the dendrimer can be tuned by number of generations and hence they can be biomimetic with natural biomolecules, especially proteins but with much higher stability. For example, G4 PAMAM (poly amidoamine dendrimer) mimics cytochrome C size and that of the G5 PAMAM mimics hemoglobin. Dendrimers exhibit antiviral activities against coronaviruses (Kandeel et al. 2020). It is evident that the structural proteins and the RNA genome of SARS CoV2 are the two major components involved in host infection and life cycle. The life cycle of SARS CoV2 clearly shows several steps involved in the host infection (Refer Chap. 3). The main step in tropism is binding to host cell. Dendrimers either prevent binding of viruses to the target cell surface or block the replication of the viral genome thereafter. S glycoprotein plays a predominant role in invading the host cell receptor, ACE2. Hence, carbohydrate-functionalized dendrimers (so-called glycodendrimers) may be a key to prevent the viral interaction with the host receptor. For example, (i) sialic acid-functionalized dendrimers (ii) carbosilane dendrimers functionalized with sialyllactose (a trisaccharide) (iii) polysulfated galactose-derivatized poly(propyleneimine) dendrimers and

(iv) mannosylated dendrimers are glycodendrimers capable of preventing host-viral interaction. Carboxylated fullerene-based dendrimers and sulfonated dendrimers are inhibitors of viral proteases which are vital enzymes for genome replication and transcription (Refer Chap. 2). Sulfonated polylysine dendrimer with a benzhydrylamine core, polyanionic dendrimers and unmodified polycationic PAMAM dendrimers are capable of inhibiting viral genome processing (Gajbhiye et al. 2009). Gene therapy using SiRNA is the promising method for gene targeting and gene silencing both in laboratory research and as clinical therapeutic owing to its high efficacy and specificity. SiRNA technology is the efficient therapeutic approach for viral diseases, including SARS CoV2. This technology however holds the demerits of hampered cell membrane penetration due to high molecular weight, rapid clearance by reticuloendothelial system and rapid glomerular filtration in the kidney and toxicity. This can however be circumvented by encapsulation or binding of the SiRNA to the dendritic nanocarriers. Dendrimers are safe non-viral vectors for siRNA carriers as they have been gradually explored since their performance in DNA delivery was approved. PAMAM dendrimers are the most investigated carriers due to ease of synthesis and modification and the commercial availability. The ciliated cells of the human lungs are the principal site for viral infection including the SARS CoV2, which can be mediated via droplets of saliva or fomites from the infected person. Thus, delivery systems designed for optimal delivery of drugs onto the lung epithelial cells can provide better and timely results. From this standpoint, PAMAM dendrimer can be a good solution. It is highly suitable for the potential aerosol-based delivery system of siRNA onto lung epithelial cells (Conti et al. 2014). ALN-RSV01 is an siRNA complementary to the N protein gene. This was the first gene-interference therapy which entered human phase II clinical trials, and this is expected to be promising against SARS CoV2 as the N protein plays a role in the life cycle of SARS CoV2 also. PAMAM dendrimers are certainly safe and effective non-viral vectors for SiRNA delivery and hold great promise for further applications in functional genomics and RNAi-based therapies. The siRNA binds to the amino terminals of G4 PAMAM and silences the genes in alveolar epithelial cells. The PAMAM-SiRNA complex can be easily administered to the virus in the aerosols also. The proposed anti-SARS efficacy of SiRNA is due to the impairment of genome replication and inhibition of viral protein synthesis in host cells (Denise et al. 2014).

Liposomes

Liposomes are bilayered nanospheres with core–shell structure formed by natural or synthetic phospholipids, mostly phosphatidylcholine constituted with hydrophilic and hydrophobic moieties. Hence, they are suitable for water-soluble and lipid-soluble drugs. They can be small unilamellar, large unilamellar or multivesicular in nature. They are also best carriers for SiRNA to initiate gene therapy by mRNA interference. They have a ball-like structural feature so that high drug and gene payload can be achieved in the cavity in addition to the bilayer membrane. They can be formulated into vaccine candidates owing to their ability to function as immunological adjuvants. Liposomal lumens are the best-known strategy for accommodating antigen cargos and for delivering without side effects or allergic reactions. Surface-bound peptide antigens exhibit viral challenging effects. Thus, liposomes are excellent adjuvant carriers in vaccine nanotechnology. In addition to neutralizing antibodies, S-proteins are also antigenic targets for cytotoxic T lymphocytes (CTL). So far, four CTL epitopes were identified from SARS CoV2 S- proteins. Yet, N protein is also an equally important antigenic target/epitope for CTL which can also elicit T cell immunity. Four peptides that were expected to be epitopes were chemically conjugated on the surface of liposomes. It was shown that two of the liposomal peptides were effective for peptide-specific CTL induction, and the most immunogenic liposomal peptide efficiently protected against viral challenge with vaccinia virus expressing this peptide. These data suggest that the surface-linked liposomal peptide may be useful for CTL-based immunotherapy against SARS CoV2. Ohno et al. have identified antigenic peptides from N proteins, considered as CTL epitopes. Two such peptides when conjugated to liposomal carriers effectively primed IFN-γ-producing CD8$^+$ T cells, elicited immune response and challenged viral infection, thus bearing a promising role as a vaccine candidate against SARS CoV2 (Ohno et al. 2009).

Respiratory endothelial cell surfaces are covered with heparan sulfate proteoglycans (HSPG). HSPG helps as an anchor for SARS CoV2 in order to contact the host cell. This can be the reason for the proposed endothelial cell dysfunction and degradation of HSPG in COVID-19 patients. Hence, HSPG may be a target for anti-SARS CoV2 drugs or vaccines. Lactoferrin is one such candidate capable of binding to HSPG and preventing the viral attachment. Thus, lactoferrin gene is upregulated in COVID-19 condition as an immune response. Lactoferrin encapsulated in phosphotidylcholine liposomes induces immunomodulatory activity (Gabriel et al. 2020; Serrano et al. 2020).

Similarly, low molecular weight molecules like heparin may also be of use in inhibiting the HSPG anchors and prevent SARS CoV2 internalization. Coronaviruses utilize common molecules such as heparan sulfate proteoglycans (HSPG) on the cell membrane to enable easy invasion into host cells. These molecules provide the first anchoring sites on the cell surface and help the virus make primary contact with host cells. LF may be able to prevent the internalization of some viruses after binding to HSPGs offering protection to the host against viral infections. LF has an important protective role in the host immune defense against COVID-19 invasion. Ivermectin has been reported to exhibit antiviral efficacy against SARS CoV2 clinical isolates in vitro. However, appropriate liposomal formulations could improve cellular internalization of ivermectin and reduce the unfavorable effect of the drug. Liposomal ivermectin possesses less cytotoxic effects and inhibits viral replication with lesser EC50 values compared to free ivermectin. Liposomal ivermectin may have a promising role in combating COVID-19. Liposomes cannot only facilitate the intracellular delivery of antivirals, and they may also encapsulate macromolecular drugs and siRNAs used to combat virus strains resistant to the currently available drugs. Ethosomes spontaneously produced by the dissolution of phosphatidylcholine and antiviral in ethanol can enhance the cell penetration. Surface-engineered liposomes prepared using stearylamine or diacetyl phosphate or mannose were reported to be ideal vesicular carrier for targeted delivery of antiviral enzyme inhibitors (Kaur et al. 2008). Surface-engineered liposomes functionalized with required ligands may be rationally designed for the specific targeting of SARS CoV2.

Nanomicelles

Nanomicelles are amphipathic spherical molecules with inner hydrophobic core for accommodating lipophilic drugs and an outer hydrophilic polymer for value-added properties such as enhanced circulatory time (Amirmahani et al. 2017). Slower rate of dissociation and consequent longer drug retention time, sustained release at the target site and eventually a higher buildup of the drug are the merits of nanomicelles which would be of enormous use in targeting viral virulent factors (Trivedi and Kompella 2010). Nanomicelles were effective in delivering antiviral agents such as curcumin (Gera et al. 2017) and such carriers will be used to deliver drugs for SARS CoV2 treatment also. Owing to the non-toxic properties and efficient antiviral activity, curcumin and nanocurcumin have been discussed under a separate topic in Chap. 5.

Solid-lipid nanoparticles (SLNs)

SLNs are colloidal lipid emulsions where the liquid lipid (oil) has been substituted by a solid lipid. Advantages of SLNs include controlled drug release, specific targeting, protection of incorporated cargo from chemical degradation, less toxicity and feasibility of large-scale production. SLN offers good encapsulation efficacy and drug release efficacy for antivirals. The SLN of monostearin functionalized with a fluorescent marker octadecylamine-fluorescein isothiocynate (ODA-FITC) prepared by solvent diffusion method in an aqueous system was identified as optimal carrier for antiviral drugs inhibiting viral attachment and impairing DNA expression (Zhang et al. 2008). SLN produced by homogenization and ultrasonication is capable of increasing the penetration of the antiviral cargo into the host cells. Bioavailability of poorly soluble antiviral drugs can be improved on encapsulation in SLN made from glyceryl monostearate and Tween 80 surfactant (Gaur et al. 2014).

Polymers

Polymeric nanoparticles have ideal properties (Table 4.2) which enhance their biomedical applications in virology. They can easily be engineered into multifunctional carriers with excellent encapsulation properties and sustained release capacity. Polymers can be useful in preparation of nanoformulations which possess relative high efficacy as compared to unformulated drug and vaccine molecules. As nanoformulation of drugs, vaccines or immunomodulators are an important strategy for treating COVID-19 patients, polymers can be used for this purpose (Itani et al. 2020). The added advantages which may be considered for the applications of polymers in the delivery of drugs and vaccines for SARS CoV2 treatment and prevention are summarized in table. Further, progressions in the manufacture of additives from commercially available antimicrobial polymers offer the possibility of rapid prototyping of a wide range of critical medical devices during the pandemic (Zuniga and Cortes 2020). Thus, there is an opportunity for the inactivation of the SARS CoV2

Table 4.2 Polymers with potential use in the management of SARS CoV2 infection

Polymer	Example	Unique feature	Potential use in virology	References
Chitosan	Cationically modified chitosan, N-(2-hydroxypropyl)-3-trimethylammonium chitosan chloride	Rapid cellular uptake Targeted delivery Live imaging	Potent inhibitor of the coronavirus entry into host	Milewska et al. (2016)
Gelatin	PEG-coated gelatin nanoparticles	Biodegradability and biocompatibility Efficient delivery of hypocrellin	IL-2 and IFN-γ release Induction of Th1 immune response Targeted drug, gene and vaccine delivery Photodynamic degradation of virus	Sahoo et al. (2015) Hirayama et al. (1997)
Dendrimer	PAMAM dendrimer	Polyvalent surface with high tenability	Intrinsic antiviral activity and improved delivery of antiviral drug to fight MERS CoV	Kandeel et al. (2020)
PLGA	PLGA nanoparticles	High encapsulation efficiency, high cellular uptake	Efficient targeted drug and vaccine delivery for respiratory viruses	Mansoor et al. (2015) Dhakal et al. (2017)

virus on critical medical devices containing functional antiviral polymers (Jorge et al. 2020).

Chitosan, a chitin derivative is a natural second most abundant biopolymer predominant in the exoskeleton of lobsters, shrimps and crabs. Chitosan encapsulated antiviral drugs are found to be more efficient than free antiviral drugs. **N-(2-hydroxypropyl) -3-trimethylammonium chitosan chloride** (HTCC) inhibits SARS CoV2 in vitro and ex vivo depending on the concentration of the chitosan and degree of substitution. HTCC is capable of interacting with DPP4 receptor (Milewska et al. 2016). Poly(ethylene glycol)-block-poly(lactide-co-glycolide) nanoparticles (PEG-PLGA nanoparticles) encapsulated with diphyllin can effectively target the ATPase enzyme and prevent the endosomal acidification. This is the way of blocking the pH-dependent internalization of coronavirus into the host cell (Hu et al. 2020).

4.3.2 Carbon Compounds

Carbon-based nanomaterials encompass zero-dimensional materials (carbon dots, nanodiamond and fullerene), one-dimensional materials like carbon nanotubes (CNTs) and two-dimensional graphene sheets or graphene oxide (GO) sheets. Superior mechanical and optoelectronic properties are common for these carbon nanomaterials. Green synthesized carbon dots have bright luminescence, aqueous stability and low cytotoxicity and therefore highly appropriate for the biomedical applications such as imaging and detection (Kim et al. 2014).

Hydrothermal carbonization of bagasse generates carbon quantum dots with high monodispersity, photoluminescence, superior photostability and aqueous dispersibility. Likewise, they show moral biocompatibility and rapid cellular internalization through the biological membrane. Hence, they can be used in biolabelling and bioimaging of cells in SARS CoV2 studies also (Du et al. 2014).

Nanodiamonds (NDs) may be in the form of diamond nanocrystals of 1–150 nm or as ultra-nanocrystalline diamond particles of 2–10 nm or in the form of diamondoids with 1–2 nm size. NDs are biocompatible and accept high payload of drugs, genes and immunomodulators, and they can easily penetrate into the membranes of cellular targets. In the diagnostic applications for COVID-19 pandemic, the high refractive index and the Raman optical activity of NDs can be exploited for cellular imaging (Kaur and Badea 2013). Multifunctionalized single-walled carbon nanotubes (SWCNTs) are ideal carriers for the sustained targeted delivery of drugs, proteins and genes. Also, SWCNTs can be a potential imaging tool in the detection of viral and host biomolecules.

Graphene oxide (GO) sheets are capable of capturing enveloped viruses, denaturing viral surface proteins and extracting the viral RNA in aqueous media. Hence, GO can be used as an antiviral and disinfectant strategy. It can also be used to capture viral proteins for the detection application. GO sheets have the potential to inhibit porcine epidemic diarrhea virus (PEDV) which is also a coronavirus like SARS CoV2. (Song et al. 2015). The sharp edges of GO nanosheets were reported to induce mechanical perturbation of structures of RNA viruses and prevent its usual binding effect to the target cell. Hence, GO may be an effective anti-SARS CoV2 material (Ye et al. 2015). The antiviral effects of GO can even be enhanced by functionalization. For example, curcumin-functionalized GO exerts antiviral efficacy by direct viral inactivation, by blocking viral attachment to the host cell and by interrupting viral replication (Yang et al. 2017). GO-silver nanocomposite is effective against RNA virus such as feline coronavirus (FCoV) which is an enveloped RNA virus similar to SARS CoV2. The antiviral efficacy is attributed to the destruction of the viral envelope and rupture of the lipid membrane. (Chen et al. 2016a, b).

Fullerenes (buckyballs) are hollow spheres of hexagonally linked carbon rings. Buckminsterfullerene (C60) is a type of fullerene with 60 carbon atoms, arranged as 12 pentagons and 20 hexagons, thus mimicking a soccer ball. As mentioned under proteomics, proteases are important antiviral targets for many viruses including SARs CoV2. The HIV-1 protease inhibitor nelfinavir was reported to strongly inhibit replication of the SARS CoV-infected vero cells and inhibited the expression of viral antigens. Nelfinavir exerted its antiviral effects in the post-entry stage of the viral life cycle, and it was recommended for designing anti-SARS drugs (Yamamoto 2004). Thus, the HIV protease inhibitors are potential SARS CoV2 inhibitors too. Of note, C60 was also a strong inhibitor of proteases in HIV virus. Model building studies and experimental verification studies show that C60 binds to the active site of proteases. The diamino C60 (the second generation C60) establishes salt bridges and Van der Waals forces with the active site of the proteases. Taken together, one can expect C60 to inhibit proteases of SARS CoV2 and design an anti-COVID-19 drug (Friedman et al. 1993). Dendrimeric derivative of C60 functionalized with carboxyl groups showed favorable interactions with the active site of viral proteases model associated with negative intermolecular energy. The ball structure fits to the active site pocket and the dendritic branches protrude outward (Schuster et al. 2000). Amino acid derivative of fullerene is an inhibitor of RdRp of cytomegalovirus. RdRP is the vital enzyme in viral gene expression. Many antivirals targeting RdRp active sites are reported to be promising SARS CoV2 agents. Hence, amino acid derivatives of fullerene can also be a potent antiviral for targeting RdRp of SARS CoV2 (Mashino et al. 2005).

Curcumin-based carbon dots impair different stages of viral life cycle in coronavirus model. The curcumin-based carbon dots act by following mechanisms:

- Prevents viral entry by impairing the structure of surface protein
- Blocks the synthesis of negative-strand RNA
- Prevents the budding of the virus
- Inhibits viral replication by activating the interferon-stimulating genes (ISGs) and stimulating the production of pro-inflammatory cytokines.

Thus, curcumin-based carbon dots possess amplified antiviral activity (Lin et al. 2019), and it can very well interfere with the life cycle of SARS CoV2 also.

4.3.3   Inorganic Nanoparticles

Hyperinflammatory state marked by fulminant cytokine storm (so-called hypercy-tokinemia) is a feature of COVID-19. As zirconia nanoparticles inhibited viral inflammation in mice model (Huo et al. 2020; Friedman et al. 1993), they can evolve as a potent anti-inflammatory agent during SARS CoV2 infection.

Silica composites exhibit virucidal effect (Balagna et al. 2020). Silicon nanoparticles have high porosity for drug and ligand loading and possess low toxicity. Silica-based nanoparticles are inexpensive and easy to prepare. They are abundant, very less toxic, highly biocompatible and also exhibit unique surface properties, such as porosity and functionalization capacity. High porosity helps for high payload for drugs and ligands, contrast agent protectors and pharmaceutical additives. Silicon nanocrystals are excellent photosensitizers of singlet oxygen (ROS) generation which results in killing of viruses as described under metal nanoparticles-induced ROS production. They can establish hydrophobic interactions with viral surface proteins. Mesoporous silicon nanoparticles are virucidal against respiratory syncytial virus (Osminkina et al. 2014). Silicon nanoparticles functionalized with glycosamino-glycans (Si-GAG) can compete with the sulfate groups of viral glycoproteins. By a competitive inhibitory mechanism, the viral binding to the host receptor can be prevented (Lee et al. 2016). As SARS CoV2 also targets the respiratory system and the infection starts with the binding of S glycoprotein to the host cell receptors, Si-GAG can be used to block host-viral interaction.

4.3.4   Metal Nanoparticles

Metal nanoparticles like gold, silver, copper, platinum and metal oxide nanoparticles such as iron oxide, zinc oxide and titanium dioxide were reported to exhibit antiviral activities. Metal nanoparticles have the ability to attack different antigen targets in the viruses with lesser probability for resistance thus proving to be a novel antiviral agent. In addition, metal nanoparticles can also be used in the detection of viruses by sensing specific proteins, peptides or nucleic acid sequences. Several workers have documented the antiviral capabilities of metal nanoparticles (Rai et al. 2016), and the advantages of metal nanoparticles which makes them excellent tools for the detection and treatment of COVID-19 are enlisted below.

- High density of active surface area
- Modifiable size and optical properties
- Facile surface chemistry
- Ability to cap phytochemicals via green synthesis method
- Ability to capture and inactivate viruses by absorbing radiation and converting into heat energy
- Selective targeting ability especially for the targeted delivery of SiRNA and subsequent viral gene knockdown.

Gold nanoparticles

The imaging and diagnosis strategies like NIR scattering, surface-enhanced Raman scattering (SERS), computed tomography (CT), magnetic resonance imaging (MRI), optical coherence tomography (OCT) and photoacoustic imaging (PAI) are used in clinical testing as well as in research for monitoring the effect of antivirals on host cell. All these techniques can be improvised by using noble metals as contrast enhancers in monitoring the viral-host interaction, virus-drug interaction, etc. Conventionally used organic dyes used as analytical probes have poor water solubility, instability and susceptibility to photobleaching, low quantum yield and less sensitivity in biological systems and weak multiplexing capability. Noble metal nanoparticles like gold and silver nanoparticles can very well avoid these disadvantages and enhance the contrast of the detection and imaging system (Moitra et al. 2020).

Gold nanoparticles (AuNPs) are best candidates for colorimetric-based biosensing applications due to their extraordinary optical properties such as high extinction coefficient, localized surface plasmon resonance (LSPR) and inherent photostability. Therefore, they are widely used in colorimetric-based biosensing of diverse targets like chemical, small molecules, proteins, metal ions and nucleic acids where the particle changes its color in response to the reactivity of the nanosized particles to the external conditions (Jazayeri et al. 2018). Gold nanoparticles have the potential to detect SARS CoV2 N proteins and S protein antibodies by giving excellent colorimetric signaling response (Moitra et al. 2020; Andresen et al. 2014). Details regarding the role of specific metal nanoparticles in sensing SARS CoV2 antigens or antibodies are discussed under Chap. 5.

Blocking the expression of a virulent gene by silencing them is called gene knockdown. Gene knockdown in coronavirus can be achieved by using small interference RNA (short interference RNA) capable of inhibiting the expression of viral mRNAS. However, short half-life and hydrolytic cleavage by RNAses are the disadvantages against the superiority of gene knockdown strategy. In this context, metal nanoparticles are designed for carrying genes by covalent and/or non-covalent mode for subsequent efficient gene delivery. The ability of metal nanoparticles to host SiRNA and other oligonucleotides can also be used in the design of RNA-based SARS CoV2 detection system with high stability and specificity (Guo et al. 2010).

Metallic nanoparticles such as copper, silver, gold and platinum are capable of generating free radicals or reactive oxygen species (ROS) like hydroxyl radicals, singlet oxygen, hydrogen peroxide, etc. ROS are noxious arms used by phagocytes, lung epithelial cells and other immune cell types against viruses. As lung epithelial cells are target for SARS CoV2, metal-mediated ROS can be used as a strategy for attacking the virus. ROS can kill pathogens directly by causing oxidative damage to vital biomolecules or indirectly by stimulating pathogen elimination by various non-oxidative mechanisms, including pattern recognition receptor signaling, autophagy, neutrophil extracellular trap formation and T-lymphocyte responses (Paiva and Bozza 2014). Thus, metal nanoparticles can be used to generate ROS and target viral cells (Fujimori et al. 2012).

Homology modeling, and in silico docking studies show that zanamivir (Hu and Ji 2020) and oseltamivir (Muralidharan et al. 2020) are efficient inhibitors of SARS CoV2. These drugs were reported to exert their antiviral mechanism by inhibiting the protease activity. However, there are lots of possibilities for the drug resistance while using these drugs (Wen et al. 2009). This issue can however be solved by gold nanoparticles coated with sialic acid. Porcine respiratory syndrome virus (PRSV), a representative model of RNA virus. Glutathione-stabilized gold nanoclusters directly inhibit the adsorption of PRSV (Bai et al. 2018), opening an avenue for targeting RNA viruses like SARS CoV2.

Silver Nanoparticles

Commercialization of silver nanoparticles-based products is constantly increasing due to their inherent antiviral properties and their ability to challenge resistant strains (Rai et al. 2014). Silver nanoparticles have proven to be active against several types of viruses including human immunodeficiency virus, hepatitis B virus, herpes simplex virus, respiratory syncytial virus, monkeypox virus and influenza viruses (Zhang et al. 2016). The mechanism of action includes interference with the crucial life cycle phases like viral attachment and penetration into host cell, viral gene replication and translation of essential proteins, intracellular viral assembly, package and gemmation (Akbarzadeh et al. 2017). Silver nanoparticles exert their antiviral effects (i) by binding to surface glycoproteins of RNA viruses thereby preventing the fusion with host cells; (ii) by reducing pro-inflammatory cytokines such as IL-6, TNF-α, CCL5 and IFNs, thereby eliciting immunomodulatory action in lung cells, both in vitro and in vivo (Park et al. 2014; Dorothea et al. 2019). Green synthesized silver nanoparticles obtained using Panax ginseng root aqueous extract were found to exhibit antiviral activity against RNA virus, suggesting the potential for attacking SARS CoV2 which is also RNA virus (Sreekanth et al. 2018). GO sheets containing AgNPs are capable of impeding the infection caused by both non-enveloped and enveloped viruses. GO sheets, which were negatively charged, interact with the positively charged lipid membranes and silver particles bind to the sulfur groups of the viral proteins. Glutathione-capped Ag_2S nanoclusters inhibit coronavirus proliferation through the blockage of viral RNA synthesis, prevention of budding and activation of the IFN-stimulating genes (ISGs) and the expression of pro-inflammation cytokines to inhibit the PEDV infection. Like PEDV, SARS CoV2 belongs to the coronavirus family suggesting that the silver-based nanosclusters may emerge as a therapy for COVID-19 (Ting et al. 2018a, b). Silver nanoparticles and silver nanowires have the ability to kill viruses causing respiratory infection (Chen et al. 2016a, b). Thus, silver nanoparticles are a promising opportunity for novel anti-SARS CoV2 which is also considerable due to the lower possibility of resistance (Galdiero et al. 2011). Following are the proposed antiviral mechanism of silver nanoparticles which shows that silver nanoparticles influence different stages of virus life cycle (Khandelwal et al. 2014) which is common for all viruses:

(1) Direct interaction with viral surface
(2) Inhibition of viral protein-host receptor interaction

Titania and Zinc Nanoparticles

Titania nanocolloids prepared by sonochemical method were reported to exhibit antiviral activity (Akhtar et al. 2019). Zinc cations acts as an antiviral by following mechanisms and holds promise in respiratory tract infection caused by SARS CoV2 (Skalny et al. 2020).

- Inhibition of SARS CoV RNA polymerase
- Decreasing the activity of SARS CoV2 receptor, ACE2
- Improving antiviral immunity by up-regulation of interferon α production
- Inhibiting inflammation and cytokine storm through the inhibition of NF-κb signaling and modulation of regulatory T cells.

The risk factors for severe COVID-19 such as aging, immune deficiency, obesity, diabetes, and atherosclerosis are associated with zinc deficiency. Hence, apart from direct antiviral action, zinc supplement may also indirectly prevent the risk of COVID-19. Intense experimental and clinical studies may result in zinc-based therapy for SARS CoV2.

4.3.5 Quantum Dots

Quantum dots (QDs) are zero-dimensional semiconductor nanoparticles or nanocrystals with diameter less than 10 nm. They have superior optical properties of super bright fluorescence, excellent photostability, narrow emission spectra and optional colors. Hence, they are traceable under specific wavelength. QDs provide multiple color imaging with a several-fold increase in brightness, thus emerging as a better fluorescent label for cellular imaging. Near-infrared emitting quantum dots are very appropriate for in vivo uses as they penetrate into deeper tissues, due to their higher wavelength as compared to visible light. Furthermore, water and the hemoglobin molecules (the major medium in vivo) do not considerably absorb these rays, thereby reducing the casual false positive signals. Despite their uses, toxicity testing and the assessment of the safety level of the quantum dots warrant further study, especially while imaging deeper layers of tissues. As quantum dots can be tuned to different sizes, specific size can be achieved in order to be compatible with the size of the SARS CoV2 (70–90 nm) and enhance penetration. Lipid-biotin conjugates have the ability to identify and mark viral lipid membranes, while streptavidin-QD conjugates were used to fluoresce them up. Thus, quantum dots are used to image enveloped viruses to be labeled in 2 h with specificity and efficiency up to 99% and 98%, respectively. This type of imaging is used to understand the complex infection mechanisms, which may be used to design diagnostic tools and therapeutics (Zhang et al. 2020). Hence, quantum dots may be used in the study of SARS CoV2, which is also an enveloped virus.

Curcumin-based cationic carbon quantum dots being positively charged are capable of binding to the S protein of coronaviruses (Ting et al. 2018a, b). Quantum dots can be used as fluorescent probes in the detection of various nanoscale processes at molecular level (Singh Ashish et al. 2017). The fluorescent property of quantum dot is equally good in virology as it reveals virus-host interaction, entry into host cells and nuclear import with essentially high temporal and spatial resolution. These multitude of information can be explored as antiviral therapeutic target. For example, quantum dot is used in mechanistic exploration of influenza viral ribonucleoprotein trafficking and host-virus interaction (Qin et al. 2019; Yamauchi 2019).

CdSe/CdS/ZnS QD coated with latex and antibodies showed photoluminescence properties in the presence of bovine serum albumin and UV irradiation and has been used in the sensitive detection of RNA virus which is also applicable for SARS CoV2 which is also a RNA virus. Rapid fluorescent immunochromatographic test (FICT) employing QD conjugate (QD-FICT) was used in the diagnosis of influenza and H1N1 virus which may be promising use in the diagnosis of SARS CoV2 (Nguyen et al. 2020). As mentioned in Chaps. 1 and 2, SARS CoV2 is characterized by an extraordinary mutation rate in their genome, relatively a million times more than their hosts! Recently, 13 variation sites were characterized in SARS CoV2 ORF1ab, S, ORF3a, ORF8 and N regions. Gene mutations are imperative in clinical diagnosis. The fluorescent property of quantum dots can be used in the detection of viral gene mutation in patient's serum. Hence, quantum dots can be used in the diagnosis of SARS CoV2 via the detection of mutation. The target DNA is amplified and hybridized with its complementary sequence which is prelabeled with biotin. Streptavidin-coated 525-nm-emission quantum dots interact with the hybrid and fluoresce. The intensity of fluorescence depends on the viral load (Zhang et al. 2017). Cadmium telluride quantum dots (CdTe QDs) were tested for the antiviral activity in pseudovirus model. The CdTe QDs have the ability to alter the confirmations of viral surface protein antigen and prevent the entry into host cells. In addition, the cadmium ions released from the QDs are capable of decreasing the viral titer in the infected cells. The antiviral effect is higher for QDs with positive functions on the surface than the negative charged QDs (Du et al. 2015).

Carbon quantum dots with phenyl boronic acid derivative, propargyl alcohol derivatized carbon quantum dots, 4-aminophenylboronic acid hydrochloride derivative of quantum dots and amino-functionalized carbon quantum dots were all strong inhibitors of human coronavirus-infected cells (HCoV-229E). The mechanism of action is the prevention of S protein-host receptor interaction and viral replication (Loczechin et al. 2019).

4.4   Functionalization of Nanoparticles

Nanostructures can be functionalized with ligands including drugs, vaccines, targeting agents, stabilizing ligands, imaging agents, anti-immunogenic agents and biocompatibility agents. Such nanostructures with two or more attachments are called multiliganded nanoparticles which function as effective therapeutic, vaccine and diagnosis tools. Multiliganded nanoparticles are very useful in antiviral strategies as it claims lesser drug intake frequency, lesser dosage, high efficacy and bioavailability

Table 4.3 Ligands used for engineering multifunctional nanoparticles

Ligand type	Examples	Applications
Stabilizing ligand	Poly ethylene glycol (PEG) Carboxylic group Sulphonic acid group Trioctylphosphine oxide (TOPO) Poly(ethylene imine) (PEI)	PEGylated interferons for antiviral activity against coronavirus
Targeting ligand	Avidin–biotin system	Serological assay of SARS CoV2 characterized by competition of antibodies and spike protein to bind to ACE2 receptor
	Monoclonal antibodies	Human anti-SARS CoV2 antibodies cross-neutralize with glycan-containing epitopes of SARS CoV2, enabling therapeutic target and detection specificity
Small molecules	Lectin	Inhibits glycan residues in spike protein
	Folate	Inhibits furin protease involved in spike protein cleavage
	Hepcidin	Inhibits ferroportin iron export receptor

(Milovanovic et al. 2017) Hence, it is essential to get introduced to different types of ligands, which are usually named based upon their function (Table 4.3).

4.4.1 Stabilizing Ligand

Stabilizing ligands are chemical groups or phytochemicals which increase the dispersity, decrease the aggregation and prevent the nanostructured cargos from immune attack such as macrophage engulfment. Stabilizing ligands are very important for maintaining the drug integrity and offering high efficient therapy. Stabilizing ligands can be engineered onto the surface or can be capped in situ as in the case of green synthesis of nanoparticles. Polar moieties like carboxylic, sulphonic acid groups, non-polar chemical groups like trioctylphosphine oxide (TOPO), triphenylphosphine (TPP), dodecanethiol (DDT), tetraoctylammonium bromide (TOAB), oleic acid (OA) neutral molecules like poly(ethylene imine) (PEI), thiol group and PEG chains are some examples of stabilizing ligands. These molecules exert repulsive forces due to electrostatic repulsion, steric exclusion or the hydration layer on the surface serving to prevent aggregation and enhance stability. Stabilizing ligand is an important factor in the drug delivery application because stability directly influences the therapeutic efficacy. Among the stabilizing ligands, the amphiphilic polymer PEG (poly(ethylene oxide or polyoxyethylene) is widely preferred as a ligand for drug delivery because of various advantages like (i) extraordinary stability (ii) high biocompatibility (iii) non-toxicity, (iv) enhanced steric effects which increases circulatory half-life (v) ability to increase the aqueous solubility of the cargo (vi)ability to escape host immune attack (vii) resistance to enzyme attack and (viii) lesser probability for renal glomerular filtration. The process of the coupling of PEG to a carrier molecule is called PEGylation, and the carriers are referred as PEGylated carriers. PEGylation is usually done by using linkers or conjugating compounds like cyanuric chloride, succinimidyl succinate, imidazoyl formate, succinimidyl carbonates or succinimidyl esters in order to link to the drug carrier. PEG can be directly linked to the cargos (proteins, peptides, enzymes, nucleic acids) or to the drug-containing carriers like polymers, micelles, nanoparticles, etc. A very good example for the first case is PEGylated cytokines. In this category, PEGylated interferons are used in the treatment of hepatitis viruses and are confirmed in vitro for attacking SARS CoV2 (Figueroa et al. 2019; Prokunina-Olsson et al. 2020; El-Lababidi et al. 2020; Sperling and Parak 2010).

4.4.2 Bioconjugation of Targeting Ligands

Bioconjugation refers to the attachment of biomolecules as ligands by chemisorption, electrostatic attraction, covalent bonding and specific molecular affinity (e.g., avidin–biotin system). Lipids, vitamins, peptides, sugars and biopolymers such as proteins, enzymes, DNA and RNA molecules or dyes can be used as bioconjugates. Bioconjugation can result in innovative properties and help in biomolecular recognition during sensing. Bioconjugation is certainly an important aspect in the SARS CoV2 management modalities. For example, avidin–biotin interaction is used in the serological assay of SARS CoV2 in order to determine the antibodies which compete for ACE2 binding. Thus, bioconjugate is used in the detection of COVID-19 (Byrnes et al. 2020). As mentioned earlier, lectin conjugation is a useful targeting strategy for virus or host cell interaction. Avidin–biotin chemistry is used for the conjugation of plant lectins to liposomes. This assembly is considered as an ideal carrier for the improved delivery of cargos to the alveolar epithelium (Bruck et al. 2001) which is infected in SARS CoV2. Peptides designed to target the interaction between the RDB of SARS CoV2 and ACE2 as well as the dimerization of ACE2 monomers can be used to target the SARS CoV2-host interaction and gain efficient therapy (Molina et al. 2020). Human monoclonal antibodies raised against SARS CoV can cross-neutralize with specific sequences in the glycan-containing epitopes of SARS CoV2 (Poh et al. 2020; Pinto et al. 2020). Aptamers designed to be complementary to the sequences pertaining to the binding sites of ACE2 or RBD domain have potential applications in probing and recognition of SARS CoV2 and also in the treatment of SARS CoV2 (Song et al. 2020; Yaling et al. 2020). Site-specific binding of the drug delivery/imaging system is enabled by targeting ligands. This helps in targeting the diseased or infected cells amidst healthy cells. For example, human monoclonal antibody that cross-neutralizes SARS CoV2 epitope can be used as a targeting ligand for the delivery of COVID-19 drug (Wang et al. 2020).

4.4.3 Small Molecules as Ligands

Small molecules are low molecular weight receptor inhibitors used for exploiting the therapeutic targets and revealing the specific biological interactions. Hence, small molecules have immense use in biomolecular recognition and in the development of diagnostic sensors. Lectin, hepcidin and folate are small molecules which may interfere with virus tropism at different stages. Structures of small molecule ligands are shown in Fig. 4.2.

Lectins

Lectins are proteins with high affinity for sugar residues. Lectins are predominant in legume seeds, roots, tubers, bulbs and rhizomes. Plant lectins can be used to specifically target the glycan residues in the glycoprotein epitopes of viruses. There are two strategies for lectin-based targeting: First, the direct targeting in which the glyconanoparticle hybrid containing the nanoparticle conjugated to the drug is glycosylated. The glycan residues can then target the cellular lectin. Secondly, lectinized nanoparticles are formed by conjugating the lectin to the nanoparticles. This assembly is then allowed to target the cellular glycan residues. This strategy is called reverse-lectin targeting. Reverse targeting using liposomes or SLNs are used for targeted drug delivery. Lung is the target for SARS CoV2; in this context, the plant lectin-bound liposomal vesicles are good at targeting alveolar epithelial cells as they show good bioadhesion and rapid cellular uptake (Bruck et al. 2001).

Folate

Folate, the water-soluble vitamin (vitamin B9) or folacin ($C_{19}H_{19}N_7O_6$) can indirectly target the S1-S2 interface of the SARS CoV2 spike protein by binding to the furin convertase and inhibit them. As furin convertase is a protease involved in the direct cleavage of S1-S2 domains, folate can be a good targeting ligand for preventing spike binding to host cells and enhance antiviral drug efficacy (Sheybani et al. 2020).

Hepcidin and hepcidin antagonists

Iron is stored in tissue as ferritin. Stored iron can be transported to the target with the help of a transport protein called transferrin. Transferrin is a glycoprotein with two subunits each of which can bind to one iron in ferric form. Transferrin enters the cell via its receptor or called ferroportin (iron gate). Iron plays a vital role in viral infections, virulence enhancement and pro-oxidant reactions in the lungs. The bronchoalveolar lavage fluid from acute respiratory distress syndrome patients (similar to that in SARS CoV2 infection) is characterized by high iron content. Hepcidin (a small peptide with 25 amino acid residues) is an inhibitor of ferroportin, which impairs iron export and elevates tissue ferritin level. In fact, the C terminus of the spike glycoprotein tail of SARS CoV2 is similar to the hepcidin. Thus, the tissue ferritin overload is one of the features of SARS CoV2 infection, which is a negative prognostic factor. Hepcidin can bind to SARS CoV2 receptors and hence prevent the host cell from infection. Conversely, anti-hepcidin molecules or hepcidin antagonists or ferroportin agonists may block the SARS CoV2's hepcidin-like motif and

Fig. 4.2 Structure of small molecule ligands used in biomolecular recognition. **a** Plant lectin (garlic lectin) an inhibitor of glycan residues in spike protein **b** folate, an inhibitor of furin protease involved in spike protein cleavage; **c**: Hepcidin, an inhibitor of ferroportin receptor

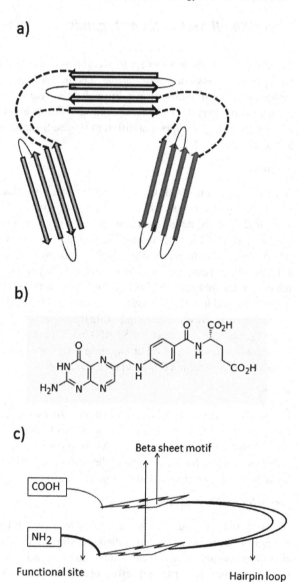

prevent the virus attachment and also promote iron export from the lungs. These two concepts can open a new therapeutic avenue in COVID-19 management. (Yilmaz and Esin 2020).

References

Abraham J. Passive antibody therapy in COVID-19. Nat Rev Immunol. 2020;20:401–3.

Ahmed EM, Solyman SM, Mohamed N, Boseila AA, Hanora A. Antiviral activity of Ribavirin nanoparticles against measles virus. Cell Mol Biol (Noisy-Le-Grand). 2018;64(9):24–32.

Akbarzadeh A, Kafshdooz L, Razban Z, Tbrizi AD, Rasoulpour S, Khalilov R. An overview application of silver nanoparticles in inhibition of herpes simplex virus. Artif Cells Nanomed Biotechnol. 2017;46(2):263–7.

Akhtar S, Shahzad K, Mushtaq S, Ali I, Rafe HM Fazal-ul-Karim SM. Antibacterial and antiviral potential of colloidal Titanium dioxide (TiO2) nanoparticles suitable for biological applications. Mater Res Express. 2019;6(10):105409

Amirmahani N, Mahmoodi NO, Glangash MM, Ghavidast A. Advances in nanomicelles for sustained drug delivery. J Indus Eng Chem. 2017;55:21–34.

Andresen H, Mager M, Griebner M, Charchar P, Todorova N, Bell N, Theocharidis G, Bertazzo S, Yarovsky I, Stevens MM. Single-step homogeneous immunoassays utilizing epitope-tagged gold nanoparticles: on the mechanism, feasibility, and limitations. Chem Mater. 2014;26(16):4696–704.

Baez-Santos YM, St John SE, Mesecar AD. The SARS-coronavirus papain-like protease: structure, function and inhibition by designed antiviral compounds. Antiviral Res. 2015;115:21–38.

Bai Y, Zhou Y, Liu H, Fang L, Liang J, Xiao S. Glutathione-stabilized fluorescent gold nanoclusters vary in their influences on the proliferation of pseudorabies virus and porcine reproductive and respiratory syndrome virus. ACS Appl Nano Mater. 2018;1(2):969–76.

Balagna C, Perero S, Percivalle EV, Ferraris M. Virucidal effect against coronavirus SARS-CoV-2 of a silver nanocluster/silica composite sputtered coating. Open Ceramics. 2020;1:100006.

Boles M, Ling D, Hyeon T. et al. The surface science of nanocrystals. Nat Mater. 2016;(15):141–153.

Bhavana V, Thakor P, Singh SB, Mehra NK. COVID-19: Pathophysiology, treatment options, nanotechnology approaches, and research agenda to combating the SARS-CoV2 pandemic. Life Sci. 2020;261:118336.

Borkow G, Zhou SS, Page T, Gabbay J. A novel anti-influenza copper oxide containing respiratory face mask. PLoS ONE. 2010;5(6):1–9.

Broglie J, Alston B, Yang C, Ma L, Adcock AF, Cheng W, Yang L. Antiviral activity of gold/copper sulfide core/shell nanoparticles against human norovirus virus-like particles. PLoS ONE. 2015;10(10):1–14.

Bruck A, Abu-Dahab R, Borchard G, Schäfer UF, Lehr CM. Lectin-functionalized liposomes for pulmonary drug delivery: interaction with human alveolar epithelial cells. J Drug Target. 2001;9(4):241–51.

Byrnes JR, Zhou XX, Lui I, Elledge SK, Glasgow JE, Lim SA, Loudermilk R, Chiu CY, Wilson MR, Leung KK, Wells JA. A SARS-CoV-2 serological assay to determine the presence of blocking antibodies that compete for human ACE2 binding. 2020;medRxiv. doi: https://doi.org/10.1101/2020.05.27.20114652.

Caly L, Druce JD, Catton MG, Jans DA, Wagstaff KM. The FDA-approved drug ivermectin inhibits the replication of SARS-CoV-2 in vitro. Antiviral Res. 2020;178:104787.

Chan WCW. Nano research for COVID-19. ACS Nano. 2020;14(4):3719–20.

Chauhan G, Madou MJ, Kalra S, Chopra V, Ghosh D, Martinez-Chapa SO. Nanotechnology for COVID-19: therapeutics and vaccine research. ACS Nano. 2020;14(7):7760–82.

Chen HW, Huang CY, Lin SY, Fang ZS, Hsu CH, Lin JC, Chen YI, Yao By, Hu CMJ. Synthetic virus-like particles prepared via protein corona formation enable effective vaccination in an avian model of coronavirus infection. Biomaterials. 2016;106:111–118.

Chen YN, Hsueh YH, Hsieh CT, Tzou DY, Chang PL. Antiviral activity of graphene–silver nanocomposites against non-enveloped and enveloped viruses. Int J Environ Res Publ Health. 2016;13(4):430.

Coleman CM, Venkataraman T, Liu YV, Glenn GM, Smith GE, Flyer DC, Frieman MB. MERS-CoV spike nanoparticles protect mice from MERS-CoV infection. Vaccine. 2017;35(12):1586–9.

Conti DS, Brewer D, Grashik J, Avasarala S, da Rocha SR. Poly(amidoamine) dendrimer nanocarriers and their aerosol formulations for siRNA delivery to the lung epithelium. Mol Pharm. 2014;11(6):1808–22.

Dhakal S, Hiremath J, Bondra K, Lakshmanappa YS, Shyu DL, Ouyang K, Kang K, Binjawadagi B, Goodman J, Tabynov K, Krakowka S, Narasimhan B, Lee CW, Renukaradhya GJ. Biodegradable nanoparticle delivery of inactivated swine influenza virus vaccine provides heterologous cell-mediated immune response in pigs. J Control Release. 2017;247:194–205.

Dorothea M, Maria A, Janice S, Teodora I, Yue Q, Antonella C, Roberto G. antiviral and immunomodulatory activity of silver nanoparticles in experimental RSV infection. Viruses. 2019;11(8):732–732.

Du F, Zhang M, Li X, Li J, Jiang X, Li Z, Hua Y, Shao G, Jin J, Shao Q, Zhou M, Gong A. Economical and green synthesis of bagasse-derived fluorescent carbon dots for biomedical applications. Nanotechnology. 2014;25(31):315702.

Du T, Cai KM, Han HY, Fang LR, Liang JG, Xiao SB. Probing the interactions of CdTe quantum dots with pseudorabies virus. Sci Rep. 2015;5:16403.

El-Atab N, Qaiser N, Badghaish H, Shaikh SF, Hussain MM. flexible nanoporous template for the design and development of reusable anti-COVID-19 hydrophobic face masks. ACS Nano. 2020;14(6):7659–65.

El-Lababidi RM, Mooty M, Bonilla MF, Salem NM. Treatment of severe pneumonia due to COVID-19 with PEG-interferon alfa 2a. ID Cases. 2020;21:e00837.

Figueroa SM, Veser A, Abstiens K, Fleischmann D, Beck S, Goepferich A. Influenza A virus mimetic nanoparticles trigger selective cell uptake. Proc Natl Acad Sci. 2019;116(20):9831–6.

Friedman SH, DeCamp D, Sijbesma RP, Srdanov G, Wudl F, Kenyon G. Inhibition of the HIV-1 protease by fullerene derivatives: model building studies and experimental verification. J Am Chem Soc. 1993;115(15):6506–9.

Frieman M, Yount B, Heise M, Kopecky-Bromberg SA, Palese P, Baric RS. Severe acute respiratory syndrome coronavirus ORF6 antagonizes STAT1 function by sequestering nuclear import factors on the rough endoplasmic reticulum/Golgi membrane. J Virol. 2007;81(18):9812–24.

Fujimori Y, Sato T, Hayata T, Nagao T, Nakayama M, Nakayama T, Sugamata R, Suzuki K. Novel antiviral characteristics of nanosized copper(I) iodide particles showing inactivation activity against 2009 pandemic H1N1 influenza virus. Appl Environ Microbiol. 2012;78(4):951–5.

Gajbhiye V, Palanirajan VK, Tekade RK. Dendrimers as therapeutic agents: a systematic review. J Pharm Pharm. 2009;61:989–1003.

Galdiero S, Falanga A, Vitiello M, Cantisani M, Marra V, Galdiero M. Silver nanoparticles as potential antiviral agents. Molecules. 2011;16(10):8894–918.

Gaur PK, Mishra S, Bajpai M, Mishra A. Enhanced oral bioavailability of Efavirenz by solid lipid nanoparticles: in vitro drug release and pharmacokinetics studies. BioMed Res Int. 2014;Article ID 363404.

Gera M, Sharma N, Ghosh M, Huynh DL, Lee SJ, Min T, Kwon T, Jeong DK. Nanoformulations of curcumin: an emerging paradigm for improved remedial application. Oncotarget. 2017;8(39):66680–98.

Ghosh S, Firdous SM, Nath A. SiRNA could be a potential therapy for COVID-19. EXCLI Journal. 2020;19:528–31.

Gong P, He X, Wang K, Wang K, Tan W, Xie W, Wu P, Li H. Combination of functionalized nanoparticles and polymerase chain reaction-based method for SARS-CoV gene detection. J Nanosci Nanotechnol. 2008;8(1):293–300.

Guo P, Coban O, Snead NM, Trebley J, Hoeprich S, Guo S, Shu Y. Engineering RNA for targeted siRNA delivery and medical application. Adv Drug Deliv Rev. 2010;62(6):650–66.

Hall DC Jr, Ji HF. A search for medications to treat COVID-19 via in silico molecular docking models of the SARS-CoV-2 spike glycoprotein and 3CL protease. Travel Med Infect Dis. 2020;35:101646.

Hirayama J, Ikebuchi K, Abe H, Kwon KW, Ohnishi Y, Horiuchi M, Shinagawa M, Ikuta K, Kamo N, Sekiguchi S. Photoinactivation of virus infectivity by hypocrellin A. Photochem Photobiol. 1997;66(5):697–700.

Hodgkinso V, Petris MJ. Copper homeostasis at the host-pathogen interface. J Biol Chem. 2012;287:13549–55.

Hu TY, Frieman M, Wolfram J. Insights from nanomedicine into chloroquine efficacy against COVID-19. Nat Nanotech. 2020;15:247–9.

Huo C, Xiao J, Xiao K,.Zou S, Wang M, Qi P, Liu T, Hu Y. Pre-treatment with zirconia nanoparticles reduces inflammation induced by the pathogenic H5N1 influenza virus. Int J Nanomed. 2020;15:661–674.

Itani R, Tobaiqy M, Al FA. Optimizing use of theranostic nanoparticles as a life-saving strategy for treating COVID-19 patients. Theranostics. 2020;10(13):5932–42.

Jazayeri MH, Aghaie T, Avan A, Vatankhah A, Ghaffari MRS. Colorimetric detection based on gold nano particles (GNPs): an easy, fast, inexpensive, low-cost and short time method in detection of analytes (protein, DNA, and ion). Sens Biosens Res. 2018;20:1–8.

Joe YH, Park DH, Hwang J. Evaluation of Ag nanoparticle coated air filter against aerosolized virus: anti-viral efficiency with dust loading. J Hazard Mater. 2016;301:547–53.

Kampf G, Todt D, Pfaender S, Steinmann E. Persistence of coronaviruses on inanimate surfaces and their inactivation with biocidal agents. J Hosp Infect. 2020;104:246–51.

Kandeel M, Al-Taher A, Park BK, Kwon HJ, Al-Nazawi M. A pilot study of the antiviral activity of anionic and cationic polyamidoamine dendrimers against the Middle East respiratory syndrome coronavirus. J Med Virol. 2020. https://doi.org/10.1002/jmv.25928.

Kaur CD, Nahar M, Jain NK. Lymphatic targeting of zidovudine using surface-engineered liposomes. J Drug Target. 2008;16(10):798–805.

Kaur R, Badea I. Nanodiamonds as novel nanomaterials for biomedical applications: drug delivery and imaging systems. Int J Nanomed. 2013;8:203–20.

Khandelwal N, Kaur G, Kumara N, et al. Application of silver nanoparticles in viral inhibition: a new hope for antivirals. Dig J Nanomater Biostruct. 2014;9:175–86.

Kim D, Choi Y, Shin E, Jung YK, Kim BS. Sweet nanodot for biomedical imaging: carbon dot derived from xylitol. RSC Adv. 2014;4:23210–3.

Kim H, Park M, Hwang J, Kim J, Chung DR, Lee KS, Kang M. Development of label-free colorimetric assay for MERS-CoV using gold nanoparticles. ACS Sens. 2019;4(5):1306–12.

Lee BY, Behler K, Kurtoglu ME, Wynosky-Dolfi MA, Rest RF Gogotsi Y. Titanium dioxide-coated nanofibers for advanced filters. J Nanopart Res 2010;12:2511−2519.

Lee EC, Davis-Poynter N, Nguyen CTH, Peters AA, Monteith GR, Strounina E, Popat A, Ross BP. GAG mimetic functionalised solid and mesoporous silica nanoparticles as viral entry inhibitors of herpes simplex type 1 and type 2 viruses. Nanoscale. 2016;8:16192–6.

Lembo D, Cavalli R. Nanoparticulate delivery systems for antiviral drugs. Antiviral Chem Chemother. 2010;21(2):53–70.

Li H, Zhou Y, Zhang M, Wang H, Zhao Q, Liu J. Updated approaches against SARS-CoV-2. Antimicrob Agent Chemother. 2020;64(6):e00483-e520.

Lin CJ, Chang L, Chu HW, Lin HJ, Chang PC, Wang RYL, Unnikrishnan B, Mao JY, Chen SY. High amplification of the antiviral activity of curcumin through transformation into carbon quantum dots. Small. 2019;15(41):1902641.

Loczechin A, Seron K, Barras A, Giovanelli E, Belouzard S, Chen YT, Metzler-Nolte N, Boukher-roub R, Dubuisson J, Szunerits S. Functional carbon quantum dots as medical countermeasures to human coronavirus. ACS Appl Mater Interf. 2019;11:42964–74.

Lu H. Drug treatment options for the 2019-new coronavirus (2019- nCoV). BioSci Trends. 2020;14(1):69–71.

Lv X, Wang P, Bai R Cong Y, Suo S, Ren X, Chen C. Inhibitory effect of silver nanomaterials on transmissible virus-induced host cell infections. Biomaterials. 2014;35(13):4195–4203.

Mansoor F, Earley B, Cassidy JP, Markey B, Doherty S, Welsh MD. Comparing the immune response to a novel intranasal nanoparticle PLGA vaccine and a commercial BPI3V vaccine in dairy calves. BMC Vet Res. 2015;11: Article number: 220.

Mashino T, Shimotohno K, Ikegami N, Nishikawa D, Okuda K, Takahashi K, Nakamura S, Mochizuki M. Human immunodeficiency virus-reverse transcriptase inhibition and hepatitis C virus RNA-dependent RNA polymerase inhibition activities of fullerene derivatives. Bioorg Med Chem Lett. 2005;15(4):1107–9.

McKee DL, Sternberg A, Stange U, Laufer S, Naujokat C. Candidate drugs against SARS-CoV-2 and COVID-19. Pharmacol Res. 2020;157:104859.

Milewska A, Kaminski K, Ciejka J, Kosowicz K, Zeglen S, Wojarski J, Nowakowska M, Szczubiałka K, Pyrc K. HTCC: broad range inhibitor of coronavirus entry. PLoS ONE. 2016;11(6):e0156552.

Milovanovic M, Arsenijevic A, Milovanovic J, Kanjeva, T, Arsenijevic N. Nanoparticles in antiviral therapy. Antimicrob Nanoarchitectonics. 2017;383–410.

Miyamoto D, Kusagaya Y, Endo N, Sometani A, Takeo S, Suzuki T, Arima Y, Nakajima K, Suzuki Y. Thujaplicin-copper chelates inhibit replication of human influenza viruses. Antiviral Res. 1998;39:89–100.

Moitra P, Alafeef M, Dighe K, Matthew B. Frieman MB, Pan D. Selective naked-eye detection of SARS-CoV-2 mediated by N gene targeted antisense oligonucleotide capped plasmonic nanoparticles. ACS Nano. 2020;14(6):7617–7627.

Muralidharan N, Sakthivel R, Velmurugan D, Michael Gromiha M. Computational studies of drug repurposing and synergism of lopinavir, oseltamivir and ritonavir binding with SARS-CoV-2 protease against COVID-19. J Biomol Struct Dyn. 2020;1–6.

Molina R, Oliva B, Fernandez-Fuentes N. A collection of designed peptides to target SARS-CoV-2 – ACE2 interaction: PepI-Covid19 database. 2020;1:100006.

Nasrollahzadeh M, Sajjadi M, Soufi GJ, Iravani S, Varma RS. Nanomaterials and nanotechnology-associated innovations against viral infections with a focus on coronaviruses. Nanomaterials. 2020;10:1072.

Nguyen AVT, Dao TD, Trinh TTT, Choi DY, Yu ST, Park H, Yeuo SJ. Sensitive 302 detection of influenza a virus based on a CdSe/CdS/ZnS quantum dot-linked rapid 303 fluorescent immunochromatographic test. Biosens Bioelectron. 2020;155:112090.

Nikaeen G, Abbaszadeh S, Yousefinejad S. Application of nanomaterials in treatment, anti-infection and detection of coronaviruses. Nanomedicine. 2020;15:1501–12.

Ohno S, Kohyama S, Taneichi M, Moriya O, Hayashi H, Oda H, Mori M, Kobayashi A, Akatsuka T, Uchida T, Matsui M. Synthetic peptides coupled to the surface of liposomes effectively induce SARS coronavirus-specific cytotoxic T lymphocytes and viral clearance in HLA-A transgenic mice. Vaccine. 2009;27(29):3912–20.

Osminkina LA, Timoshenko VY, Shilovsky IP, Kornilaeva GV, Shevchenko SN, Gongalsky MB, Tamarov KP, Abramchuk SS, Nikiforov VN, Khaitov MR, Karamov EV. Porous silicon nanoparticles as scavengers of hazardous viruses. J Nanopart Res. 2014;16:2430–40.

Paiva CN, Bozza MT. Are reactive oxygen species always detrimental to athogens? Antioxid Redox Signal. 2014;20(6):1000–37.

Palmieri V, Papi M. Can graphene take part in the fight against COVID-19? Nano Today. 2020;33:100883.

Park S, Park HH, Kim SY, Kim SJ, Woo K, Ko G. Antiviral properties of silver nanoparticles on a magnetic hybrid colloid. Appl Environ Microbiol. 2014;80(8):2343–50.

Pinto D, Park Y, Beltramello M, Walls AC, Tortorici MA, Bianchi S, et al. Cross-neutralization of SARS-CoV-2 by a human monoclonal SARS-CoV antibody. Nature. 2020;583:290–5.

Poh CM, Carissimo G, Wang B, Amrun SN, Lee CYP, Chee RSL et al. Two linear epitopes on the SARS-CoV-2 spike protein that elicit neutralising antibodies in COVID-19 patients. Nat Commun. 2020;11: Article Number: 2806.

Prokunina-Olsson L, Alphonse N, Dickenson RE, Durbin JE, Glenn JS, Hartmann R, Kotenko SV, Lazear HM, O'Brien TR, Odendall C, Onabajo OO, Piontkivska H, Santer DM, Reich NC, Wack A, Zanoni I. COVID-19 and emerging viral infections: the case for interferon lambda. J Exp Med. 2020;217(5):e20200653.

Qin C, Li W, Li Q, Yin W, Zhang X, Zhang Z, Zhang X, Cui Z. Real-time dissection of dynamic uncoating of individual influenza viruses. Proc Nat Acad Sci USA. 2019;116:2577–82.

Rai M, Deshmukh SD, Ingle AP, Gupta IR, Galdiero M, Galdiero S. Metal nanoparticles: the protective nanoshield against virus infection. Crit Rev Microbiol. 2016;42(1):46–56.

Rai M, Kon K, Ingle A, Duran N, Galdiero S, Galdiero M. Broad-spectrum bioactivities of silver nanoparticles: the emerging trends and future prospects. Appl Microbiol Biotechnol. 2014;98:1951–61.

Roh C, Jo SK. Quantitative and sensitive detection of SARS coronavirus nucleocapsid protein using quantum dots-conjugated RNA aptamer on chip. J Chem Technol Biotechnol. 2011;86:1475–9.

Ruiz-Hitzky E, Darder M, Wicklein B, Ruiz-Garcia C, Martín-Sampedro R, Real G, Aranda P. Nanotechnology responses to COVID-19. Adv Healthc Mater. 2020;9:2000979.

Sahoo N, Sahoo RK, Biswas N, Guha A, Kuotsu K. Recent advancement of gelatin nanoparticles in drug and vaccine delivery. Int J Biol Macromol. 2015;81:317–31.

Sarkar DS. Silver nanoparticles with bronchodilators through nebulisation to treat COVID 19 patients. J Curr Med Res Opin. 2014;3(4):449–50.

Schuster DI, Wilson SR, Kirschner AN, Schinazi RF, Schluter-Wirtz S, Tharnish P, Barnett T, Ermolieff J, Tang J, Brettreich M, Hirsch A. Evaluation of the anti-HIV potency of a water-soluble dendrimeric fullerene. Proc Electrochem Soc. 2000;9:267–70.

Serrano G, Kochergina L, Albors A, Diaz E, Oroval M, Hueso G, Serrano JM. Liposomal lactoferrin as potential preventative and cure for COVID-19. Int J Res Health Sci. 2020;8(1):8–15.

Shetty R, Ghosh A, Honavar SG, Khamar P, Sethu S. Therapeutic opportunities to manage COVID-19/SARS-CoV-2 infection: present and future. Indian J Ophthalmol. 2020;68 (5):693–702.

Sheybani Z, Dokoohaki MH, Negahdaripour M, et al. The role of folic acid in the management of respiratory disease caused by COVID-19. ChemRxiv. 2020. https://doi.org/10.26434/chemrxiv.12034980.

Singh L, Kruger HG, Maguire GEM, Govender T, Parboosing R. The role of nanotechnology in the treatment of viral infections. Ther Adv Inf Dis. 2017;4(4):105–31.

Singh Ashish K, Pradyot P, Ranjana S, Nabarun N, Zeba F, Monika B, Singh Ranjan K, Anchal S, Roy Jagat K., Brahmeshwar M, Singh Rakesh K. Curcumin quantum dots mediated degradation of bacterial biofilms. Front Microbiol 2017

Sivasankarapillai VS, Pillai AM, Rahdar A, Sobha AP, Das SS. Mitropoulos AC, Mokarrar MH, Kyzas GZ. On facing the SARS-CoV-2 (COVID-19) with combination of nanomaterials and medicine: possible strategies and first challenges. Nanomaterials. 2020;10 (5):852.

Skalny AV, Rink L, Ajsuvakova OP, Aschner M, Gritsenko VA, Alekseenko S, Svistunov AA, Petrakis D, Spandidos DA, Aaseth J, Tsatsakis A, Tinkov AA. Zinc and respiratory tract infections: perspectives for COVID 19. Int J Mol Med. 2020;46:17–26.

Song ZY, Wang XW, Zhu GW, Nian QG, Zhou HY, Yang D, Qin CF, Tang RK. Virus capture and destruction by label-free graphene oxide for detection and disinfection applications. Small. 2015;11:1171–6.

Sperling RA, Parak WJ. Surface modification, functionalization and bioconjugation of colloidal inorganic nanoparticles. Philos Trans Roy Soc A. 2010;368:1333–1383.

Sportelli MC, Izzi M, Kukushkina EA, Hossain SI, Picca RA, Ditaranto N, Cioffi N. Can nanotechnology and materials science help the fight against SARS-CoV-2? Nanomaterials. 2020;10(4):802.

Sreekanth TVM, Nagajyothi PC, Muthuraman P, Enkhtaivan G,. Vattikuti SVP, Tettey CO, Kim DH, Shim J Yoo K. Ultra-sonication-assisted silver nanoparticles using panax ginseng root extract and their anti-cancer and antiviral activities. J Photochem Photobiol B Biol. 2018;188:6–11.

Sucipto TH, Churrotin S, Setyawati H, Kotaki T, Martak F, Soegijanto S. Antiviral activity of copper(II) chloride dihydrate against dengue virus type-2 in Vero cell. Indonesian J Trop Infect Dis. 2017;6(4):84–9.

Ting D, Nan Dong N, Fang L, Lu J, Bi J, Xiao S, Han H. multisite inhibitors for enteric coronavirus: antiviral cationic carbon dots based on curcumin. ACS Appl Nano Mater. 2018;1(10):5451–9.

Ting D, Liang J, Dong N, Lu J, Fu Y, Fang L, Xiao S, Han H. Glutathione-capped Ag2S nanoclusters inhibit coronavirus proliferation through blockage of viral RNA synthesis and budding. ACS Appl Mater Interf. 2018;10(5):4369–4378.

Trivedi R, Kompella UB. Nanomicellar formulations for sustained drug delivery: strategies and underlying principles. Nanomedicine. 2010;5(3):485–505.

Tse LV, Rita MM, Rachel G L, Ralph BS. The current and future state of vaccines, antivirals and gene therapies against emerging coronaviruses. Front Microbiol. 2020;11: Article 658.

Udugama B, Kadhiresan P, Kozlowski HN, Malekjahani A, Osborne M, Li VYC, Chen H, Mubareka S, Gubbay J, Chan WCW. diagnosing COVID-19: the disease and tools for detection. ACS Nano. 2020;14(4):3822–35.

Uskokovic V. Nanotechnologies: what we do not know. Technol Soc. 2007;29(1):43–61.

Van Doremalen N, Bushmaker T, Morris DH, Holbrook MG, Gamble A, Williamson BN, et al. Aerosol and surface stability of SARS-CoV-2 as compared with SARS-CoV-1. N Engl J Med. 2020;382:1564–7.

Verdecchia P, Angeli F, Reboldi G. Angiotensin-converting enzyme inhibitors, angiotensin II receptor blockers and coronavirus. J Hypertens. 2020;38(6):1190–1.

Wang C, Li W, Drabek D. Okba NMA, Haperen RV, Osterhouse ADME, van Kuppeveld FJM, Haagmans BL, Grosveld F, Bosch BJ. A human monoclonal antibody blocking SARS-CoV-2 infection. Nat Commun. 2020;11: Article No: 2251.

Warnes SL, Little ZR, Keevil CW. Human coronavirus 229E remains infectious on common touch surface materials. mBio. 2015;6 (6):e01697–15.

Wen WH, Lin M, Su CY, Wang SY, Cheng YSE, Fang JM, Wong CH. Synergistic effect of zanamivir-porphyrin conjugates on inhibition of neuraminidase and inactivation of influenza virus. J Med Chem. 2009;52:4903–10.

Wrapp D, De Vlieger D, Corbett KS, Torres GM, Wang N, Van Breedam W, Roose K, van Schie L, Hoffmann M, Pöhlmann S, et al. Structural basis for potent neutralization of betacoronaviruses by single-domain camelid antibodies. Cell. 2020;181(5):1004-1015.e15.

Yang XX, Li CM, Li YF, Wang J, Huang CZ. Synergistic antiviral effect of curcumin functionalized graphene oxide against respiratory syncytial virus infection. Nanoscale. 2017;9:16086–92.

Yamamoto N. HIV protease inhibitor nelfinavir inhibits replication of SARS-associated coronavirus. Biochem Biophys Res Commun. 2004;318(3):719–25.

Yamauchi Y. Quantum dots crack the influenza uncoating puzzle. Proc Nat Acad Sci USA. 2019;116(7):2404–6.

Yanling S, Jia S, Xinyu W, Mengjiao H, Miao S, Lin Z, et al. (2020): Discovery of aptamers targeting receptor-binding domain of the SARS-CoV-2 spike glycoprotein. Anal Chem. 2020;92(14):9895–990.

Yilmaz N, Esin E. Covid-19 and Iron Gate: The role of transferrin, transferrin receptor and hepcidin. 2020. https://www.researchgate.net/publication/340860987.

Yue H, Wei W, Fan B, Yue Z, Wang L, Ma G, Su Z. The orchestration of cellular and humoral responses is facilitated by divergent intracellular antigen trafficking in nanoparticle-based therapeutic vaccine. Pharmacol Res. 2012;65(2):189–97.

Zhang XG, Miao J, Li MW, Jiang SP, Hu FQ, Du YZ. Solid lipid nanoparticles loading adefovir dipivoxil for antiviral therapy. J Zhejiang Univ Sci B. 2008;9(6):506–10.

Zhang XF, Liu ZG, Shen W, Gurunathan S. Silver nanoparticles: synthesis, characterization, properties, applications, and therapeutic approaches. Int J Mol Sci. 2016;17(9):E1534.

Zhang C, Chen Y, Liang X, Zhang G, Ma H, Nie L, Wang Y. Detection of hepatitis B virus M204I mutation by quantum dot-labelled DNA probe. Sensors. 2017;17:961.

Zhang LJ, Wang S, Xia L, Cheng L, Tang HW, Liang Z, Xiao G, Pang DW. Lipid-specific labeling of enveloped viruses with quantum dots for single-virus tracking. mBio. 2020;11 (3):e00135–20.

Zumla A, Chan J, Azhar E. Hui DSC, Coronaviruses—drug discovery and therapeutic options. Nat Rev Drug Discov. 2016;15:327–47.

Zuniga JM, Cortes A. The role of additive manufacturing and antimicrobial polymers in the COVID-19 pandemic. Expert Rev Med Devices. 2020;17(6):477–81.

Chapter 5
Existing and Promising Methods of Diagnosis for COVID-19

Diagnostic approaches for SARS CoV2 include:

- Nucleic acid amplification through real-time polymerase chain reaction (RT-PCR) (Chakraborty et al. 2020; Udugama et al. 2020)
- Scanning methods (chest radiography and computed tomography/CT) (Chakraborty et al. 2020; Udugama et al. 2020)
- Analysis of SARS CoV2 protein epitopes (Chen et al. 2015)
- Serological assay (identification of neutralization antibodies raised against SARS CoV2) (Meyer et al. 2014)
- Biochemical assays of immune markers such as C-reactive protein, lymphocyte counts, levels of interleukins IL6 and IL 10 (Ai et al. 2020)
- Various integrated point-of-care molecular approaches and nanosensing methods which are under development (Mathuria et al. 2020).

Collection of appropriate samples from the suspected patients is very crucial for the reliable, precise and timely diagnosis of COVID-19. In addition to the conventional nasopharyngeal swab, the lower respiratory tract specimens such as sputum and bronchoalveolar lavage (BAL) may also be collected (Michael et al. 2020). According to the Center for Disease Control and Prevention (Center for Disease Control and Prevention 2020), the upper respiratory specimen, especially the nasopharyngeal specimen is preferred for the RT-PCR method of detection. Alternate choices are oropharyngeal specimen, nasal mid-turbinate swab, an anterior nares specimen and nasopharyngeal wash/aspirate or nasal aspirate specimens. Blood samples are collected for serological assay. Some research reports have suggested the reliability of blood, urine samples and anal swab as a source of SARS CoV2 detection (Peng et al. 2020). The healthcare professionals strictly adhere to the infection prevention and control guidelines while collecting the samples and storing them in viral transport medium. However, false positive results and false negative results must be carefully considered while analyzing any sample type and also while using any diagnostic method. This chapter focuses on few approved methods as well as some promising emerging tools for SARS CoV2 detection.

© The Author(s), under exclusive license to Springer Nature Singapore Pte Ltd. 2021
Devasena T., *Nanotechnology-COVID-19 Interface*, Nanotheranostics,
https://doi.org/10.1007/978-981-33-6300-7_5

5.1   Nanoparticles-Assisted RT-PCR

RT-PCR is the standard method for the detection of SARS CoV2 as this method enables the result in real time while the procedure is still continuing (the conventional RT-PCR however gives the results at the end of the protocol). Further, RT-PCR is considered as a "gold standard" as most of the molecular detection method for SARS CoV2 detection such as loop-mediated isothermal amplification and clustered regularly interspaced short palindromic repeats and multiplex isothermal amplification are based on RT-PCR assays (Tang et al. 2020). The World Health Organization has approved RT-PCR-based diagnostic testing procedure for the diagnosis of COVID-19 based on the detection of specific genome sequence of SARS CoV2 (World Health Organization (2020). Based on this approved protocol, diagnostic kits have been manufactured and supplied to public health sectors for testing (Ahn et al. 2020).

For RT-PCR assay, various molecular targets (the genes) of SARS CoV2 genome can be considered for amplification. The targets include the genes coding for structural proteins of SARS CoV2 such as S protein, E protein, M protein and N protein or the genes encoding non-structural proteins such as helicase and RdRp (Tang et al. 2020) (Refer Chap. 2). At least two targets should be assayed for the confirmed diagnosis of COVID-19. WHO recommends the detection of E gene followed by RdRp gene as first line detection for COVID-19 (Corman et al. 2020). Forward and reverse primers 5′-TCAGAATGCCAATCTCCCCAAC-3′ and 5′-AAAGGTCCACCCGATACATTGA-3′, respectively, can be used for targeting the SARS CoV2 envelope E genome (Huang et al. 2020) in combination with the previously reported probes for targeting the RdRp region of SARS CoV2 (Jung et al. 2020).

Using the primers, the target viral gene is amplified and measured by the cycle threshold (Ct), which is defined as the number of cycles required for the fluorescent signal to cross the threshold and becomes detectable. A Ct value less than 40 is clinically reported as PCR positive (Sethuraman et al. 2020). Among the different SARS CoV2 target genes mentioned above the combination of RdRp/Hel gene, RT-PCR possesses higher sensitivity and specificity (Chan et al. 2020). ORF1 gene of SARS CoV2 can be analyzed with high throughput and fast turnaround time by using the recently marketed fully automated commercial RT-PCR. This is capable of detecting larger volume of patients in a shorter time frame (Poljak et al. 2020; Corman et al. 2020). Development of COVID-19 RdRp/Hel into a multiplex assay which can simultaneously detect other human-pathogenic coronaviruses and respiratory pathogens may further increase its clinical utility in the future. (Chan et al. 2020).

Nsp2 of SARS CoV2 can also be targeted by PCR assay. RT-PCR involving the primers targeted toward the longest and previously untargeted SARS CoV2-specific nsp2 region is used for the rapid, sensitive and reproducible assay of SARS CoV2 nsp2. This method shows an imprecision of less than 5% and does not target the genes of other coronaviruses or respiratory viruses. Moreover, this method is sensitive with a limit of detection (LOD) of 1.8 $TCID_{50}$/ml (Yip et al. 2020).

Disadvantages of RT-PCR assays:

- Lengthy nucleic acid extraction procedure
- No on-site detection
- Possibility for false negative results (due to unstable reagents, low viral load in the sample and improper sampling techniques)
- Sophisticated detection method
- Needs experienced manpower
- Need complex equipment
- Needs cold system for reagents sample transport and storage
- Time consumption.

Nanoparticles-assisted RT-PCR assays are relatively superior to conventional assays due to high sensitivity, rapidity, specificity and efficacy. Hence, nano-RTPCR will be a promising tool in the diagnosis and the field surveillance of SARS CoV2 infection which will also deliver high throughput detection kits to fight with this pandemic. The nanotechnology intervention applied for the diagnosis of respiratory viruses SARS CoV and MERS CoV may be promising in the diagnosis of COVID-19, thus opening new research avenues in the area of SARS CoV2 detection. For example, incorporation of super paramagnetic nanoparticles and gold nanoparticles in PCR technique enables high sensitive detection of SARS CoV and PEDV (porcine epidemic diarrhea virus) which may be used for SARS CoV2 detection in near future (Gong et al. 2008; Yuan et al. 2015). Gold nanoparticles with surface-engineered-specific oligonucleotides result in high signal amplifications during the detection of coronaviruses like *TGEV* (transmissible gastroenteritis *virus*) and PEDV (Zhu et al. 2017). The same principle may be applied to the detection of SARS CoV2 which is also a member of coronavirus. Gold nanoparticles-assisted PCR assay was developed for detection of bovine respiratory syncytial virus with high sensitivity which may be applicable for the SARS CoV2 that targets the respiratory system (Liu et al. 2019).

5.2 Loop-Mediated Isothermal Amplification (Lamp) Assay

Loop-mediated isothermal amplification (LAMP) is an assay method to detect COVID-19 that can be an alternative for the RT-PCR. In LAMP method, the gene sequences are amplified in an isothermal condition with high speed and high specificity without the need of sophisticated equipments and trained analysts. Point-of-care device based on LAMP can be a promising COVID-19 diagnostic method (Nguyen et al. 2020). Specific high sensitivity enzymatic reporter unlocking (SHERLOCK) protocols/lateral flow assay involving clustered regularly interspaced short palindromic repeats (CRISPR) and associated Cas proteins 12 and 13 are used in rapid and accurate detection of SARS CoV2 infection using respiratory swab RNA extracts. This assay involves simultaneous reverse transcription and isothermal amplification using loop-mediated amplification (RT-LAMP) of genes encoding S protein and N protein (Xiang et al. 2020 and Broughton et al. 2020).

5.3   Serodiagnosis

Though gene detection by PCR is the currently preferred method for diagnosing COVID-19, there are chances for false negatives and also it is not feasible for asymptomatic carriers. Therefore, detecting specific anti-SARS CoV2 antibodies is an alternative choice for COVID-19 diagnosis. Immunogobulin G (IgG) an important indicator of immune response in the acute infection period can be a target for the on-site detection.

Lateral flow assay utilizing colloidal gold nanoparticles (AuNP-LF) was developed to achieve rapid diagnosis and on-site detection of the IgM antibody developed against the SARS CoV2 through the indirect immune chromatography method. The AuNP-LF strip is composed of a sample pad, a conjugate pad, a nitrocellulose membrane, an absorbent pad and all attached to a PVC backing card. For preparing AuNP-LF strips, antihuman IgM conjugated nanoparticles which functions as the detecting reporter are placed in the conjugation pad at the left end of the backing card. Next, the SARS CoV2 nucleoprotein was coated on an analytical nitrocellulose membrane for sample capture (the test line, T). Goat antimouse IgG was also immobilized on the membrane as control line (C line) next to the test line. If the test sample contains antibody against the N protein of SARS CoV2 (the IgM), it will be captured in the T line and sandwiched between N protein and the gold nanoparticle reporter. Some amount of gold reporter binds to the C line also. This would give a signal as a legible red T line. If the sample is negative for SARS CoV2 antibodies, then the reporter binds to C line only and there will not be any sandwich in the T line to give a signal (Huang et al. 2020).

Advantages of IgM detection:

- High specificity and sensitivity
- Less time consumption
- Less sample requirement
- Easy to operate
- On-site detection
- Low cost
- Highly suitable for emergency detection.

As per recent report, lanthanide nanoparticles-based lateral flow immunoassay is used for the diagnosis of COVID-19 by the detection of IgG produced by SRS CoV2 infection. The assay system contains a nitrocellulose membrane impregnated with N protein of SARS CoV2 (produced by recombinant DNA technology) which is used for capturing the complementary IgG. Anti-human IgG (raised in mouse) labeled with self-assembled lanthanide-doped polystyrene nanoparticles functions as a fluorescent reporter. This assay can monitor the progression of COVID-19, as well as responses to SARS CoV2 treatment (Chen et al. 2020).

An immunochromatographic assay (ICA) using immobilized gold nanoparticles capable of probing the spike protein of PEDV can be extrapolated for detecting SARS CoV2 as both are coronaviruses. This method uses a strip containing a sample pad, conjugation pad, nitrocellulose membrane and absorption pad lying one after the other. Antispike protein antibodies conjugated to gold nanoparticles are adsorbed to the conjugation pad (at one end of the strip). The capturing antibody lies in the nitrocellulose membrane (at the other end of the strip). When SARS CoV2 positive sample is dropped on the sample pad, it reacts with the antibody-bound gold nanoparticle and gets sandwiched by the capture antibody to form a red signal, enabling the on-site detection (Bian et al. 2019).

5.4 Gold Nanoparticles-Assisted Visual Detection

Visual detection of target molecules is enabled by measuring the absorbance as specific wavelength of nanoparticles or nanoparticle-conjugated dye using a colorimeter. For example, gold nanoparticles functionalized with thiol group, oligonucleotide sequence complementary to ORF1 and sequences upstream of E protein gene of MERS CoV genome can be considered. The hybrid establishes covalent bond with the gold nanoparticles and prevents salt-induced aggregation leading to transition of optical properties and enabling visual detection. This method has a very low limit of detection. Gold nanohybrid-based method can be used for diagnosis of other infectious diseases including SARS CoV2 (Kim et al. 2019).

Gold nanoaparticles fabricated with poly-dimethylsiloxane (PDMSO) is bound to gold-binding polypeptide. The polypeptide functions as a fusion partner and enables nanopatterning of proteins including SARS CoV antigen. SARS CoV E protein and green fluorescent protein when patterned on gold nanoparticles and allowed to interact with complementary antibodies may produce color change which can be quantitatively used for SARS CoV detection (Park et al. 2006). Owing to sequence homology, SARS CoV2 E protein can also be detected using this method.

COVID-19 can be detected by sensing the N protein gene using gold nanoparticles-based colorimetry. Gold nanoparticles were coated with the thiol-modified antisense oligonucleotides (ASOs) (approximately 30 nm) specific for N-gene. In the presence of N protein gene, the ASOs will form a hybrid which would change the surface plasmon resonance of gold nanoparticles. Absorbance can be measured at 660 nm. A hyperchromic shift indicates the binding and aggregation of target gene. Confirmation of the assay can be done using RNASeH. In the presence of RNAseH, the target DNA is hydrolyzed so that no change, i.e., red shift will be observed. This method holds promising selectivity in the presence of MERS CoV. This method is also sensitive with a detection limit of 0.18 ng/μL. This method may be applicable for SARS CoV2 detection (Moitra et al. 2020).

Immunogenic targets are of utmost importance in the development of specific detection tools and vaccine candidates. Two linear epitopes on the SARS CoV2 spike protein that elicit neutralizing antibodies were identified in human COVID-19 patients, and they are IgG immunodominant regions (Poh et al. 2020). Such epitopes when fabricated into a sensor by binding to nanoparticles can be used to induce detectable changes in an assay system. For example, gold nanoparticles surface engineered with the short viral peptide epitopes are used in the detection of viral antibodies based on immunoassay. In the presence of viral infection, the antibodies will specifically bind to the epitope and result in the aggregation of gold nanoparticles and producing a detectable colorimetric signal (Andersen et al. 2014).

5.5 Nanoimmunosensor

Graphene-based field-effect transistor (GFET) is developed into a nanoim-munosensor which is capable of sensing SARS CoV2 in human nasopharyngeal swab. GFET is coated with antibodies raised against SARS CoV2 spike protein via 1-pyrenebutyric acid N-hydroxysuccinimide ester (PBASE) linker. The S protein antigen in the sample binds to the antibodies and produces electrochemical signal with high specificity (Seo et al. 2020).

5.6 RNA Aptamers-Based Optical Detection

RNA aptamer (modified with amine at the 5 'end) designed for specific binding to SARS CoV N protein is conjugated to quantum dots (with carboxyl group) through EDC coupling. The aptamer hybridizes with the N protein pre-immobilized in a glass chip and produces signal with high sensitivity. Quantum dots are responsible for producing the optical signal. The signal intensity can be imaged using confocal laser scanning microscope. Detection of multiple antigens can be done using the aptamers specific for the desired target, suggesting the possibility for the detection of COVID-19 using SARS CoV2 N protein-specific aptamer (Roh et al. 2011).

5.7 Point-of-Care Detection by Gold Nanoparticles

Detection of suspected samples without transporting them to centralized facilities is called point-of-care approach. Point-of-care detection of COVID-19 can be carried out by lateral flow of the SARS CoV2 antigen in a test strip. The sample (containing the antigen) is dropped on a double-streaked membrane strip consisting of antibody-conjugated gold nanoparticle (wine red color) in one streak and capture antibodies in the other streak. The sample moves laterally by capillary flow and binds to the gold nanoparticle antibody, and subsequently it gets immobilized by the captured antibodies forming a cluster of blue color owing to the coupling of the plasmon band (Xiang et al. 2020).

5.8 Biobarcode Assay

In biobarcode assay, magnetic nanoparticles were functionalized with capture anti-bodies that are complementary to one of the epitope of the target protein. Simultaneously, gold nanoparticles were functionalized with detection antibodies capable of binding to another epitope of the same antigen. In addition to the antibody, the gold nanoparticles contain biobarcodes (nucleic acid sequences). The antigen gets sandwiched between the magnetic nanoparticle and the gold nanoparticle and forms a complex which can be isolated by magnetic field and the biobarcode removed by adding dithiothreitol. The biobarcodes are bound to the complementary sequences immobilized on glass chip and then detected by fluorophore or silver-based scattering enhancement. Optically tunable nanoparticle probes have been used for the ultra-sensitive detection of specific proteins and nucleic acid sequences including the viral genome (Bao et al. 2006; Hill et al. 2006; Thanxton et al. 2009). Gold nanoparticles and quantum dots are often used in biobarcode assay for viral detection (Kim et al. 2016). This technique holds promise in detecting SARS CoV2 proteins and nucleic acid sequences.

5.9 Protein Nanobubbles and Metal Nanoparticles as Contrast Enhancers for In vivo Imaging

According to Kanne et al., chest radiography and CT imaging findings are the vital supportive tools of COVID-19 diagnosis for radiologists (Kanne et al. 2020). Chest X-ray radiography projected in the anterioposterior projection is considered as a reliable and easy tool for detecting SARS CoV2 infection in emergency setting. The most common X-ray pattern is multifocal and peripheral, associated with interstitial and alveolar opacities. The parenchymal abnormalities detectable by X-ray imaging are as follows: (1) alveolar opacities (hazy increase in lung attenuation with no obscuration of the underlying vessels) (2) interstitial opacities or thickening (3) alveolar opacities associated with consolidations and (4) pleural effusions (Ippolito et al. 2020).

Gold nanoparticles can enhance the contrast of X-ray images as compared to typical iodine-based contrast agents owing to higher atomic number and X-ray absorption coefficient (Hainfeld et al. 2006). Gold nanoparticles covered with a bilayer of polyelectrolyte when injected will be engulfed by cells wherein they scatter incident × radiation in detectable amounts, giving an enhanced signal as compared to cell without gold. Gold nanoparticle-enhanced X-ray scatter images are more sensitive than typical absorption-based X-ray imaging (Rand et al. 2011).

Spatial frequency heterodyne imaging (SFHI) is a novel X-ray scatter imaging technique which makes use of nanoparticle contrast agents. The high sensitivity of SFHI as compared to traditional absorption-based X-ray radiography and the resulting anisotropic information may be utilized for the detection of lungs infected with SARS CoV2. Iron oxide nanoparticles and gold nanoparticles have the ability to enhance the image quality of SFHI (Stein et al. 2010). Novel apoferritin "nanobubble" contrast agents filled with perfluoropropane (C_3F_8) gas are proven to be versatile in vivo image enhancers in SFHI (Rand et al. 2014). SFHI integrated with nanostructures will be of immense use in lung imaging during SARS CoV2 infection.

CT imaging discloses the bilateral lung involvement (which correlates with disease stage, patient age and immune status) while the thin slice CT scan images reveal the thickening of the interlobular septa. High-resolution CT (HRCT) reveals small, honeycomb-like condensation of the interlobular septa. CT imaging is used to identify different changes in different stages of COVID-19. The early stage features (1–3 days after the emergence of clinical signs and symptoms) include single or numerous patchy GGOs segregated by grid-like condensed or honeycomb-like interlobular septa. The rapid progression stage features (3–7 days after the emergence of clinical signs and symptoms) include fibrous exudation attached to every alveolus throughout the interalveolar space, creating a fusion situation. The consolidation stage features (6–15 days after the appearance of clinical signs and symptoms) include numerous patchy pulmonary consolidations of lower density. The dissipation stage (14th–21st day) features include strip-like opacity and patchy consolidation (Jin et al. 2020). Intravenous injection of gold nanoparticles of 30 nm size enables in vivo computed tomography (CT) imaging with high contrast and without background images. This is due to easy separation of scattered radiations from transmitted ones as they reach the detector from different angles (Kim et al. 2007).

References

Ahn DG, Shin HJ, Kim MH, Lee S, Kim HS, Myoung J, Kim BT, Kim SJ. Current status of epidemiology, diagnosis, therapeutics, and vaccines for novel coronavirus disease 2019 (COVID-19). J Microbiol Biotechnol. 2020; 30(3):313–24.

Ai T, Yang Z, Hou H, Zhan C, Che, C, Lv W, Tao Q, Sun Z, Xia L. Correlation of chest CT and RT-PCR testing in coronavirus disease 2019 (COVID19) in China: a report of 1014 cases. Radiology. 2020; 200642.

Andresen H, Mager M, Grießner M, Charchar P, Todorova N, Bell N, Theocharidis G, Bertazzo S, Yarovsky I, Stevens MM. Single-step homogeneous immunoassays utilizing epitope-tagged gold nanoparticles: on the mechanism, feasibility, and limitations. Chem Mater. 2014;26(16):4696–704.

Bao YP, Wei TF, Lefebvre PA, An H, Kunkel GT, Muller UR. Detection of protein analytes via nanoparticle-based bio bar code technology. Anal Chem. 2006;78:2055–9.

Bian H, Xu F, Jia Y, Wang L, Deng S, Jia A, Tang Y. A new immunochromatographic assay for on-site detection of porcine epidemic diarrhea virus based on monoclonal antibodies prepared by using cell surface fluorescence immunosorbent assay. BMC Veterinary Res. 2019;15(1):32.

Broughton JP, Deng X, Yu G, Fasching CL, Singh J, Streithorst J, Granados A, Sotomayor-Gonzalez A, Zorn K, Gopez A, Hsu E, Gu W, Miller S, Pa, CY, Guevara H, Wadford DA, Chen JS, Chiu CY. Rapid detection of 2019 novel coronavirus SARS-CoV-2 using a CRISPR-based DETECTR lateral flow assay. 2020; https://doi.org/10.1101/2020.03.06.20032334.

Centre of Disease Control and Prevention. Coronavirus disease 2019 (COVID-19)—guideline for clinical specimen. 2020; https://www.cdc.gov/coronavirus/2019-ncov/lab/guidelines-clinical-specimens.html.

Chakraborty C, Sharma AR, Sharma G, Bhattacharya M, Lee SS. Sars-CoV-2 causing pneumonia-associated respiratory disorder (COVID-19):diagnostic and proposed therapeutic options. Eur Rev Med Pharmacol Sci. 2020;24:4016–26.

Chan JF, Yip CC, To KK, Tang THC, Wang SCY, Leung KH, et al. Improved molecular diagnosis of COVID-19 by the novel, highly sensitive and specific COVID-19-RdRp/Hel real-time reverse transcription-PCR assay validated in vitro and with clinical specimens. J Clin Microbiol. 2020;58(5):e00310-e320.

Chen Y, Chan KH, Kang Y, Chen H, Luk HK, Poon RW. A sensitive and specific antigen detection assay for middle east respiratory syndrome coronavirus. Emerg Microbes Infect. 2015;4:e26.

Chen Z, Zhang Z, Zhai X, Li Y, Lin L, Zhao H, Bian L, et al. Rapid and sensitive detection of anti-SARS-CoV-2 IgG using lanthanide-doped nanoparticles-based lateral flow immunoassay. Anal Chem. 2020;92:7226–31.

Corman VM, Landt O, Kaiser M, Molenkamp R, Meijer A, Chu DKW, Bleicker T, Brünink S, Schneider J, Schmidt ML, Mulders DGJC, Haagmans BL, van der Veer B, van den Brink S, Wijsman L, Goderski G, Romette JL, Ellis J, Zambon M, Peiris M, Goossens H, Reusken C, Koopmans MPG, Drosten C. Detection of 2019 novel coronavirus (2019-nCoV) by real-time RT-PCR. Eurosurveillance. 2020;25(3):2000045.

Gong P, He X, Wang K, et al. Combination of functionalized nanoparticles and polymerase chain reaction-based method for SARS-CoV gene detection. J Nanosci Nanotechnol. 2008;8(1):293–300.

Hainfeld JF, Slatkin DN, Focella TM, Smilowitz HM. Gold nanoparticles: a new X-ray contrast agent. Br J Radiol. 2006;79(939):248–53.

Hill HD, Mirkin CA. The bio-barcode assay for the detection of protein and nucleic acid targets using DTT-induced ligand exchange. Nat Protoc. 2006;1(1):324–36.

Huang C, Wang Y, Li X, Ren L, Zhao J, Hu Y, Zhang L, Fan G, Xu J, Gu X, Cheng Z. Clinical features of patients infected with 2019 novel coronavirus in Wuhan China. Lancet. 2020;395:497–506.

Huang C, WenT, Shi FJ, Zeng XY, Jiao YJ. Rapid detection of IgM antibodies against the SARS-CoV-2 virus via colloidal gold nanoparticle-based lateral-flow assay. ACS Omega. 2020; 5(21):12550–6.

Ippolito D, Maino C, Pecorelli A, Allegranza P, Cangiotti C, Capodaglio C, Mariani I, Giandola T, Gandola D, Bianco I, Ragusi M, Franzesi CT, Corso R, Sironi S. Chest X-ray features of SARS-CoV-2 in the emergency department: a multicenter experience from northern Italian hospitals. Respir Med. 2020;170:106036.

Jin YH, Cai L, Cheng ZS, Cheng H, Deng T, Fan YP, Fang C, Huang D, Huang LQ, Huang Q, Han Y, Hu B, Hu F, Li BH, Li YR, Liang K, Lin LK, Luo LS, Ma J, Ma LL, Peng ZY, Pan YB, Pan ZY, Ren XQ, Sun HM, Wang Y, Wang YY, Weng H, Wei CJ, Wu DF, Xia J, Xiong Y, Xu HB, Yao XM, Yuan YF, Ye TS, Zhang XC, Zhang YW, Zhang YG, Zhang HM, Zhao Y, Zhao MJ, Zi H, Zeng XT, Wang YY, Wang XH. A rapid advice guideline for the diagnosis and treatment of 2019 novel coronavirus (2019-nCoV) infected pneumonia. Military Med Res. 2020;7:4.

Jung YJ, Park GS, Moon JH, Ku K, Beak SH, Kim S, Park EC, Park D, Lee JH, Byeon CW, Lee JJ, Maeng SJ, Kim SI, Kim B, Kim MJ, Kim S, Kim BT, Lee MG, Kim HG. Comparative analysis of primer-probe sets for the laboratory confirmation of SARS-CoV-2 BioRxiv. 2020; https://doi.org/10.1101/2020.02.25.964775.

Kanne JP, Chest CT. Findings in 2019 novel coronavirus (2019-nCoV) infections from Wuhan, China: key points for the radiologist. Radiology. 2020;295:16–7.

Kim D, Park S, Lee JH, Jeong YY, Jon S. Antibiofouling polymer-coated gold nanoparticles as a contrast agent for in vivo X-ray computed tomography imaging. J Am Chem Soc. 2007;129(24):7661–5.

Kim H, Park M, Hwang J, Kim JH, Chung DR, Lee KS, Kang M. Development of label-free colorimetric assay for MERS-CoV using gold nanoparticles. ACS Sensor. 2019;4(5):1306–12.

Kim J, Biondi MJ, Feld JJ, Chan WCW. Clinical validation of quantum dot barcode. ACS Nano. 2016;10(4):4742–53.

Liu Z, Li J, Liu Z, et al. Development of a nanoparticle-assisted PCR assay for detection of bovine respiratory syncytial virus. BMC Veterinary Res. 2019;15:110.

Mathuria JP, Yadav R, Rajkumar. Laboratory diagnosis of SARS-CoV-2—a review of current methods. J Infect Publ Health. 2020; 13(7):901–5.

Meyer B, Drosten C, Muller MA. Serological assays for emerging coronaviruses: challenges and pitfalls. Virus Res. 2014;194:175–83.

Michael LJ, Yi-Wei T. Laboratory diagnosis of emerging human coronavirus infections—the state of the art. Emerg Microbes Infect 2020; 9:747–56.

Moitra P, Alafeef M, Dighe K, Frieman MB, Pan D. Selective naked-eye detection of SARS-CoV-2 mediated by N gene targeted antisense oligonucleotide capped plasmonic nanoparticles. ACS Nano. 2020;14(6):7617–27.

Nguyen T, Duong Bang D, Wolff A. Novel coronavirus disease (COVID-19): paving the road for rapid detection and point-of-Care diagnostics. Micromachines. 2020;11(3):306.

Park TJ, Lee SY, Lee SJ Park JP, Yang KS, Lee KB. Protein nanopatterns and biosensors using gold binding polypeptide as a fusion partner. Anal Chem. 2006; 78(20):7197–205.

Peng L, Liu J, Xu W, Luo Q, Chen D, Lei Z. SARS-CoV-2 can be detected in urine, blood, anal swabs, and oropharyngeal swabs specimens. J Med Virol. 2020. https://doi.org/10.1002/jmv.25936.

Poh CM, Carissimo G, Wang B, Amrun SN, Lee SYP, Chee RSL et al. Two linear epitopes on the SARS-CoV-2 spike protein that elicit neutralising antibodies in COVID-19 patients. Nature Commun. 2020;11:2806.

Poljak M, Korva M, Knap Gašper N, et al. Clinical evaluation of the cobas SARSCoV-2 test and a diagnostic platform switch during 48 hours in the midst of the COVID-19 pandemic. J Clin Microbiol. 2020. https://doi.org/10.1128/JCM.00599-20.

Rand D, Ortiz V, Liu Y, Derdak Z, Wands JR, Tatíček M, Rose-Petruck C. Nanomaterials for X-ray imaging: gold nanoparticle enhancement of X-ray scatter imaging of hepatocellular carcinoma. Nano Lett. 2011;11(7):2678–83.

Rand D, Uchida M, Douglas T, Rose-Petruck C. X-ray spatial frequency heterodyne imaging of protein-based nanobubble contrast agents. Opt Express. 2014;22(19):23290–8.

Roh C, Jo SK. Quantitative and sensitive detection of SARS coronavirus nucleocapsid protein using quantum dots-conjugated RNA aptamer on chip. J Chem Technol Biotechnol. 2011;86:1475–9.

Seo G, Lee G, Kim MJ, et al. Rapid detection of COVID-19 causative virus (SARS-CoV-2) in human nasopharyngeal swab specimens using field-effect transistor-based biosensor. ACS Nano. 2020;14(4):5135–42.

Sethuraman N, Jeremiah SS, Ryo A. Interpreting diagnostic tests for SARS-CoV-2. JAMA. 2020;323(22):2249–51.

Stein AF, Ilavsky J, Kopace R, Bennett EE, Wen H. Selective imaging of nanoparticle contrast agents by a single-shot x-ray diffraction technique. Opt Express. 2010;18(12):13271–8.

Tang YW, Schmitz JE, Persing DH, Stratton CW. The laboratory diagnosis of COVID-19 infection: current issues and challenges. J Clin Microbiol. 2020; https://doi.org/10.1128/JCM.00512-20.

Thaxton CS, Elghanian R, Thomas AD, Stoeva S, Lee I, Smith JS, Schaeffer ND, Klocker AJ, H, Horninger W, Bartsch G. Nanoparticle-Based bio-barcode assay redefines "undetectable" PSA and biochemical recurrence after radical prostatectomy. Proc Nat Acad Sci U.S.A. 2009; 106(44):18437–42.

Udugama B, Kadhiresan P, Kozlowski HN, Malekjahani A, Osborne, Li VYC et al. Diagnosing COVID-19: The disease and tools for detection. ACS Nano. 2020; 14(4):3822–35.

World Health Organization. (2020). Laboratory testing of 2019 novel coronavirus (2019-nCoV) in suspected human cases: interim guidance, 17 January 2020. World Health Organization. https://apps.who.int/iris/handle/10665/330676.

Xiang J, Yan M. Evaluation of enzyme-linked immunoassay and colloidal gold-immunochromatographic assay kit for detection of novel coronavirus (SARS-Cov-2) causing an outbreak of pneumonia (COVID-19). 2020. https://doi.org/10.1101/2020.02.27.20028787.

Xiang X, Qian K, Zhang Z. CRISPR-cas systems based molecular diagnostic tool for infectious diseases and emerging 2019 novel coronavirus (COVID-19) pneumonia. J Drug Target 2020; 1–5. https://doi.org/10.1080/1061186X.2020.1769637.

Yip CC, Ho CC, Chan JF, To KKW, Ying Chan HS, Ying Wong SC, et al. Development of a novel, genome subtraction-derived, SARS-CoV-2-specific COVID-19-nsp2 real-time RT-PCR assay and its evaluation using clinical specimens. Int J Mol Sci. 2020;21(7):2574.

Yuan W, Li Y, Li P, Song Q, Li L, Sun J. Development of a nanoparticle-assisted PCR assay for detection of porcine epidemic diarrhea virus. J Virol Methods. 2015;220:18–20.

Zhu Y, Liang L, Luo Y et al. A sensitive duplex nanoparticle-assisted PCR assay for identifying porcine epidemic diarrhea virus and porcine transmissible gastroenteritis virus from clinical specimens. Virus Genes. 53(1):71–6.

Chapter 6
Potential Therapeutic Approaches for SARS CoV2 Infection

6.1 The Four Target Points

Many therapeutic approaches are promising in the treatment of SARS CoV2 infection. Yet, all these approaches discussed in this chapter come under any one of the broad groups of strategies (denoted as target points I, II, III and IV). Compounds, drugs or techniques involved in this strategy are illustrated in Table 6.1.

I.   Targeting viral entry into the host cytoplasm
II. Targeting viral gene replication in host cells
III. Targeting the viral proteins (enzymes, structural proteins and nsps) expressed in the host cells.
IV. Inhibiting exaggerated host immune response.

I.   **Targeting viral entry to host cells**:
     Convalescent plasma therapy, antibodies (Abraham 2020) and nanobodies (Wrapp et al. 2020) are used to prevent viral attachment and entry into host cells. The antimalarial drug chloroquine (McKee et al. 2020) follows several mechanisms to kill SARS CoV2, and the first mechanism is inhibition of viral entry. Antiviral drug molecules are also inhibitors of viral attachment (Verdecchia et al. 2020; Shetty et al. 2020).
II. **Targeting viral gene replication in host cells**:
     The antimalarial drug chloroquine (McKee et al. 2020), siRNA (Ghosh et al. 2020), replication enzyme inhibitors (Caly et al. 2020) and transcription enzyme inhibitors (Zumla et al. 2016) could be therapeutic options that can block SARS CoV2 gene expression.
III. **Targeting the viral proteins (enzymes and structural proteins)**:
     Proteins expressed in the host cells, nuclear import pathway inhibitors (Frieman et al. 2007) and inhibitors of enzyme activities of the non-structural proteins (nsp inhibitors) come under this category.
IV. **Inhibiting the exaggerated host immune response**:

Devasena T., *Nanotechnology-COVID-19 Interface*, Nanotheranostics, https://doi.org/10.1007/978-981-33-6300-7_6

Table 6.1 List of potential therapeutics used to treat SARS CoV2 infection

Category	Example	Mechanism of action	Point of target
Antibodies	Convalescent plasma (neutralizing antibodies)	Neutralizing the surface antigens of SARS CoV2	I
	CR3022		
	47D11		
	Anti-TMPRSS2 antibodies		
	Nanobodies		
Antimalarial drug	Chloroquine	Binds to ACE2 receptor. Binds to sialic acid and prevents acidification. Inhibits cathepsin-mediated endosomal maturation	I
	Chloroquine	Prevents the genome replication in coronaviruses	II
Gene therapy	SiRNA	Knocks down the target genes and blocks the synthesis of proteins and enzymes essential for viral replication, protein synthesis and survival mechanism	II, III
	Ivermectin	Inhibits nuclear transport	III
3CLpro inhibitors	Paritaprevir and rateglavir	Inhibits SARS CoV2 genome replication	II
PLpro inhibitors	Formoterol	Inhibits polyprotein cleavage and non-structural protein formation (refer Chap. 2) Increases the host immunity and decreases the immune evasion capacity of the SARS CoV2	III
2′OMTase inhibitors	Dolutegravir and bictegravir	Interferes with mRNA capping and decreases its stability and subsequent translation machinery	II, III
NTPase inhibitors	Arylketoacids	Inhibits SARS CoV2 genome expression	II
Helicase inhibitors	Potassium bismuth citrate	Inhibits SARS CoV2 genome replication	II

(continued)

Table 6.1 (continued)

Category	Example	Mechanism of action	Point of target
RdRp inhibitors	Ribavirin Aurin tricarboxylic acid (ATA)	Inhibits SARS CoV2 genome polymerization and replication	II
TMPRSS2 inhibitors	Camostat mesylate	Inhibits spike protein fusion to host cell Suppresses the activity of cytokines interleukin-1beta (IL-1b) and interleukin-6 (IL-6) and controls inflammation	I, III, IV
ACE2 inhibitors	Arbidol	Blocks S glycoprotein and inhibits virus–host ACE2 binding	I
Viroporin inhibitors	Salinomycin	Inhibits proton channel matrix protein and prevents endosomal acidification	I
Heparan sulfate inhibitors	Dendritic polyglycerol Tellurium nanostars with bovine serum albumin	Prevents viral binding	I
DPP4 inhibitor	Sitagliptin	Inhibits viral attachment and fusion	
Lipid raft inhibitors	Cyclodextrin	Inhibits viral attachment and entry	I
SKP2 inhibitors	Niclosamide	Terminates viral genome replication	II
STAT3 inhibitor	Silibinin	Inhibits viral replication	II
Inhibitor of cytokine storm	Leukemia-inhibiting factor Silibinin	Inhibits alveolar damage and promotes regenerative repair Inhibits cytokine storm	IV
Inhibitor of lung damage	SiRNA candidates: SiSC1 and SiSC5	Inhibits alveolar damage	IV
Lectin binder	Mannose and galactose residues	Inhibits ACE2-independent infection pathway	I

Anti-inflammatory agents, leukemia inhibition factor (LIF), immunomodulators and mitogen = activated phosphokinase inhibitors are used for ameliorating the effects of exaggerated host immune response and cytokine storm associated with SARS CoV2 infection.

Different compounds with potential for treating SARS CoV2 infection are shown in Fig. 6.1. and discussed below:

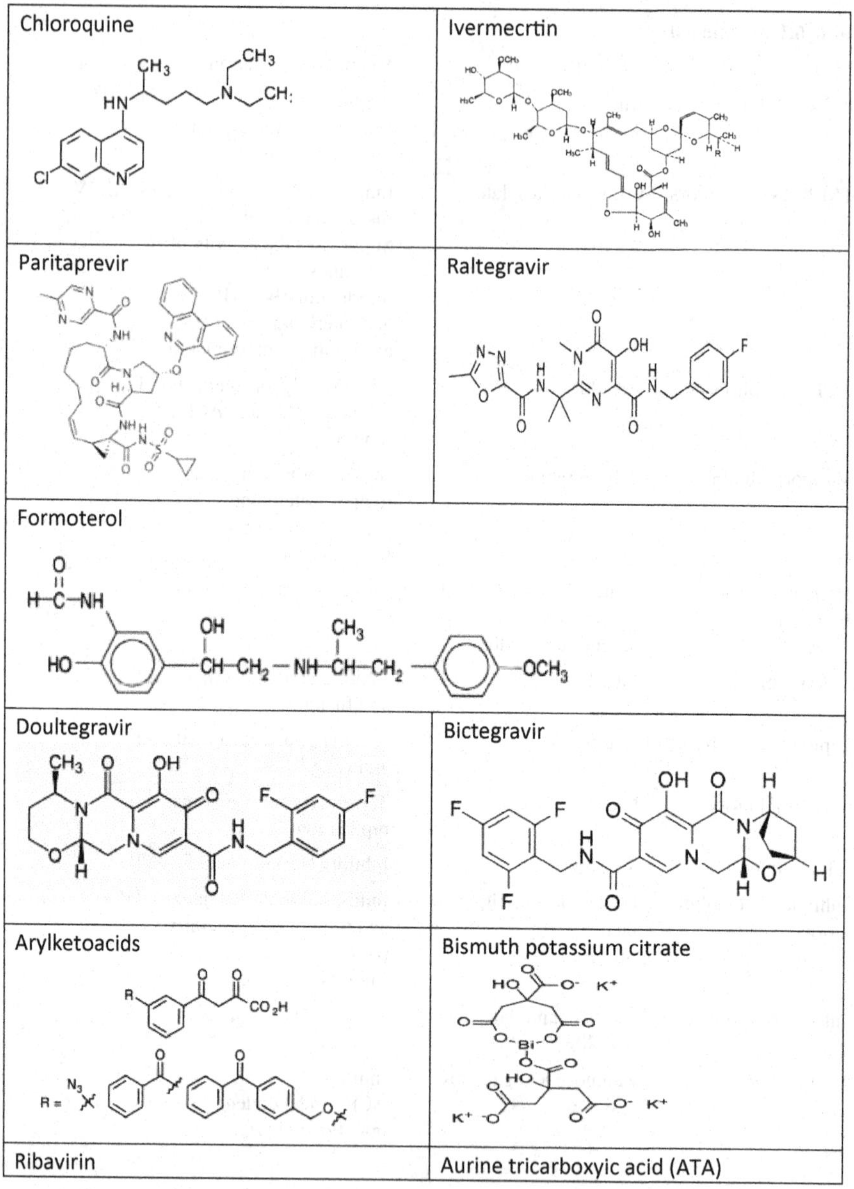

Chloroquine	Ivermecrtin

Paritaprevir	Raltegravir

Formoterol

Doultegravir	Bictegravir

Arylketoacids	Bismuth potassium citrate

Ribavirin	Aurine tricarboxyic acid (ATA)

Fig. 6.1 Structures of therapeutic compounds used to combat SARS CoV2 infection

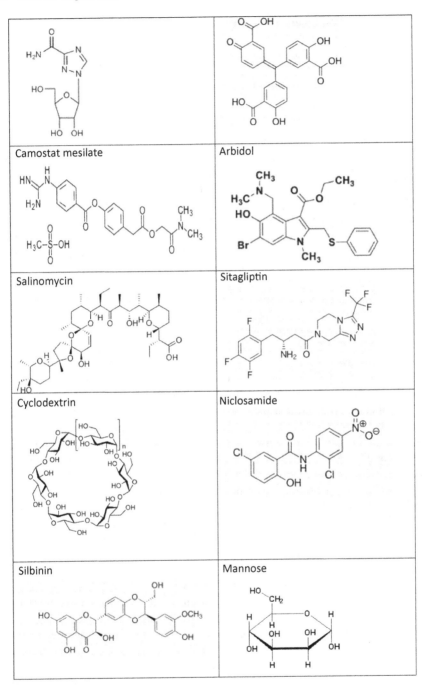

Camostat mesilate | Arbidol

Salinomycin | Sitagliptin

Cyclodextrin | Niclosamide

Silbinin | Mannose

Fig. 6.1 (continued)

6.2 Antibody-Based Therapy

Antibody-based therapy for SARS CoV2 infection can be carried out by 3 strategies:

- Convalescent plasma transfusion therapy (CPTT)
- Monoclonal antibodies
- Nanobodies.

Convalescent plasma transfusion therapy (CPTT) is a classic adaptive immunotherapy effectively used in the management of SARS, MERS and 2009 H1N1 pandemic with high efficacy and safety (Cheng et al. 2005; Ko et al. 2018). As far as SARS CoV2 pandemic is concerned, CPTT is the most promising and successful method of treatment (Sheriden 2020). CPTT protocol is approved by the US Food and Drug Administration which recommends the consent from the patients and the donor after confirmation of COVID-19 negative results from the latter (Tanne 2020). In this method, the plasma is collected from the SARS CoV2-infected patients (so-called convalescent plasma) in the recovery state but without symptoms. The convalescent plasma is rich in neutralizing antibodies which are capable of counteracting with the SARS CoV2 antigens as well as virus–host interaction and decrease the viral infectivity in seven to ten days per dose of 200 ml (Duan 2020).

When CPTT is not feasible, monoclonal antibodies complementary to the antigens of SARS CoV2 can be synthesized by hybridoma technology and used as in CPTT. For example, anti-S glycoprotein antibodies of SARS CoV prevent the binding between the virus and the ACE2 receptor of the host cell and stop the infection (Savarino et al. 2006). CR3022 is a S1 glycoprotein-specific antibody of recombinant origin, which binds to the RBD domain of the spike protein and prevents the SARS CoV2 infection (Tian 2020). 47D11 is a human monoclonal antibody complementary to the conserved epitopes on the RBD of the spike protein for cross-neutralizing the SARS CoV2 (Wang et al. 2020a, b). The original antibody, human IgG1 was also reported to have cross-neutralizing effect on SARS CoV2, thus opening an opportunity for a potential therapeutic candidate by itself or in combination with other neutralizing antibodies (Tian et al. 2019). SARS CoV2 life cycle stages have revealed the role of TMPRSS2 serine protease in priming SARS CoV2 spike protein-mediated host entry through ACE2 receptor. Hence, it is valid to propose that antibody complementary to SARS CoV spike may inhibit SARS CoV2 infectivity too (Hoffmann et al 2020).

Thus, antibodies (in the form of CPTT, monoclonal antibodies/recombinant antibodies/human Igs) are frontliners to prevent viral entry into the host cells. Hence, it is worth enhancing the effect of antibodies using nanotechnological concepts. In this line, nanoliposomes are appropriate carriers for the targeted delivery of viral antibodies. For example, nanoliposomes (detailed in Chap. 4) are an efficient carrier for the targeted delivery of indinavir (Gagne 2002).

Limited stability of antibodies and the low level of tissue penetration are the demerits of antibodies which necessitate higher dose. Nanoencapsulation of therapeutic antibodies may assist in potential targeting, reduced immunogenicity, superior bioavailability, lesser degradation and controlled release. Besides, antibody nanoencapsulation may also enable binding to the intracellular targets including enzymes, oncogenic proteins and transcription factors (Sousa et al. 2017). The polyethylene glycol-poly(ε-caprolactone) encapsulation enhances the efficacy of antibodies-based treatment approaches (Xu et al. 2018). Conjugation of antibodies to nanopolymers enhances the antigen-binding ability, paratope availability and efficacy of antibody-based nanomedicines. Fab fragments of the antibodies can be covalently conjugated to PLGA–PEG nanoparticles using disulfide-selective pyridazinedione linkers and strain-promoted alkyne–azide click chemistry to achieve better efficacy (Greene et al. 2018). Targeting of SARS CoV2 antigens with high efficacy can thus be achieved by nanoconstructions of antibodies using liposomes and polymeric nanocapsules.

Another member of antibody family is nanobodies (Nbs) which may be a substitute for CPTT. Nbs are recombinant single-domain antibodies (approximately 20 nm) isolated from the variable VHH domains of the "heavy chain only antibodies (HCAbs) of camelids such as dromedaries, llamas, alpacas, etc. (Hamers-Casterman et al. 1993; Muyldermans 2013). Nbs can be raised against antigenic epitopes of viruses and be used as an immunotherapeutic tool, for example, Nbs of E2 glycoprotein of hepatitis C virus functions as potential candidates for immunotherapeutic administration in chronic hepatitis C (Alexander et al. 2013). Nbs elicited toward SARS CoV2 antigenic epitopes may be tested for COVID-19 treatment.

6.3 Chloroquine Has the Potential to Inhibit the Endocytosis of Viral Nanostructures

Chloroquine (Fig. 6.1) is an antimalarial drug used in the clinical application for almost 70 years! Chloroquine is effective in the treatment of the SARS CoV2 infection in vitro (Wang et al. 2020a, b). It is also effective in clinical phases to treat patients with COVID-19 at different levels of severity. Subsequently, the supremacy of chloroquine was proved in 20 clinical studies launched in several Chinese hospitals, in which the drug decreased the exacerbation of pneumonia, severity and the duration of the symptoms and rapidly cleared the viral titer without detectable side effects (Gao et al. 2020b; Colson et al. 2020). At the entry level of the SARS CoV2, chloroquine can follow many anti-infective mechanisms (Devaux et al. 2020). It binds with the ACE2 receptor and prevents the SARS CoV2 attachment to the target cells. It can also interfere with the host cell sialic acid binding and neutralizes the acidification of lysosomes, thereby blocking the dropping of endosomes into the host cytoplasm. Chloroquine is also capable of inhibiting the cathepsins which is an acid protease requiring a low pH for optimal cleavage of SARS CoV2 S1–S2 interface. This would impair the formation of the autophagosomes and subsequent viral release.

In addition to its ability to interfere with the viral attachment to the host (target point I), chloroquine possesses antireplicative ability (target point II) which can prevent the replication in coronavirus (Keyaerts et al. 2009). The dose needed for inhibiting the coronavirus replication is *on par* with the dose used for the antimalarial treatment (EC_{50} value of 8.8 ± 1.2 μM) (Keyaertse et al. 2004). Chloroquine is also capable of interfering with the mitogen-activated protein kinase (MAP kinase) pathway (target point III) and also with the anti-SARS CoV2 CD8[+] T cells to ameliorate the production of pro-inflammatory cytokines (target point IV) (Devaux et al. 2020).

SARS CoV2, being a spherical nanostructure with size range of 90–120 nm (Kim et al. 2020), is considered similar to the synthetic nanoparticles produced in the laboratory (Wolfram and Ferrari 2019; Zhu et al. 2019; Cilurzo et al. 2013). As chloroquine is useful in analyzing the interaction of cells with nanospheres (Wolfram and Ferrari 2019), it will be equally effective in the nanomedicine to understand the interaction between SARS CoV2 (so-called viral nanostructure) and the host cells! (Hu et al. 2020). Nanostructures depend on phosphatidylinositol binding clathrin assembly protein (PICALM) and clathrin-mediated endocytosis to gain entry into the cytoplasm (Wolfram et al. 2017). Hence, if PICALM is targeted, clathrin-mediated endocytosis can be prevented (Miller et al. 2015). Chloroquine is an ideal PICALM targeting agent, which can suppress the clathrin-mediated endocytosis (Hu et al. 2020). Another mechanism by which chloroquine prevents the entry of lysosomal nanosized cargos like SARS CoV2 is the enhancement of lysosomal pH which will consequently prevent autophagy pathway (Amaravadi et al. 2011; Hu et al. 2020). Thus, considering SARS CoV2 as a nanostructure, chloroquine can be utilized to study the different aspects of this virus.

Chloroquine when formulated into poly-lactic acid (PLA) nanospheres possesses feasible properties like high biodegradability, biocompatibility, sustained release and great potential to improve the antiviral activity at lower doses (Lima et al. 2018). Chloroquine–gold nanoparticles are good at stimulus-responsive release of the drug in a controlled fashion to get better efficacy without side effects (Prachi et al. 2012). Gelatin nanocarriers loaded with chloroquine exhibit high encapsulation efficiency and sustained release (Bajpai et al. 2006).

6.4 Organic and Inorganic Nanocarriers in SiRNAs Strategy

Small interfering RNAs (short interfering RNAs/siRNAs) are short oligomers designed to be complementary sequences to a target sequence. SiRNA can be fabricated to base pair with the genomic RNAs and/or sgmRNAs of the SARS CoV2 which can block the expression of the latter. As a result, the replication proteins, structural proteins and the virulent proteins encoded in the target genes will not be translated, thereby affecting the infective mechanism and crucial stages of the life cycle. SiRNAs have been used to treat SARS in humans and SARS CoV2 infection in *Macaca mulatta* model after intranasal administration. SiRNAs are effective in reducing fever, viral titer and acute diffuse alveoli damage without toxic effects (Li et al. 2005; Huang et al. 2020).

In order to completely avail the anti-infective property of siRNA, it should be secured during the delivery process especially from the engulfment by immune cells and from the RNase cleavage. For this purpose, stabilizing ligands and nanocarriers are used. PEGylated cationic liposomes encapsulating siRNA (so-called cationic lipoplexes) are highly competent in target gene knockdown without susceptibility to enzyme and immune attack. The effect also depends on the types of lipids used in the liposomal vesicle formulation (Hattori et al. 2019). Hence, liposomal delivery of siRNA may be an option for gene therapy strategy to cure SARS CoV2 infection. Nanopolymeric structures of polyaspartamide and siRNA hold promise in targeting lung cells and treating asthma (Craparo et al. 2020). This approach may be used for COVID-19 treatment, as breathing insufficiency is a characteristic feature of SARS CoV2 infection. Polyethyleneimine-modified magnetic Fe_3O_4 nanoparticles were prepared for the delivery of therapeutic siRNAs (Jin et al 2019). Carbonate apatite nanoparticles trigger pH-dependent co-release of drugs and siRNA into the target cells (Fatemian and Moghimi 2019). Thus, organic nanocarriers like liposomes, polymeric nanostructures and inorganic nanoparticles like carbonate apatite and iron oxide nanoparticles can be used in siRNA delivery strategies for the treatment of COVID-19.

6.5 Lipid Carriers for Ivermectin

Ivermectin (Fig. 6.1) is a macrocyclic lactone derived from *Streptomyces avermitilis* initially used for its antiparasitic activity. Now ivermectin is reported to exhibit antiviral activity against SARS CoV2 (Heidary and Gharebaghi 2020). It is known to inhibit importin (IMPα/β1)-mediated nuclear import of SARS CoV2 nuclear protein in clinical isolate samples (Yang et al. 2020; Caly et al. 2020).

Ivermectin molecules however show the following disadvantages (Lu et al. 2017; Silva et al. 2016; Balaña-Fouce et al. 1998) which can very well be troubleshot by nanotechnology: (i) first-pass effect as they are prone to enzyme attack and inactivation as evidenced in mammalian models; (ii) poor membrane penetration accompanied by low bioavailability and lesser therapeutic effects; and (iii) drug resistance by target cells which mandates multiple doses for better results. Nanocarriers can increase the efficacy of ivermectin-based COVID-19 treatment. Sustained and targeted release of ivermectin can be achieved by encapsulation into lipid nanocarriers. Nanoencapsulated cargos possess high antiviral efficacy as compared to free ivermectin (Ahmadpour et al. 2019). SLNs prepared by hot homogenization and ultrasonication approach show high encapsulation efficiency for amorphous ivermectin and put on extended release without burst discharge (Dou 2016; Guo et al. 2018). Liposomes made from soy lecithin and cholesterol show less toxicity and are less resisted by the target cells (Velebny et al. 2000). Nanosuspensions of ivermectin show high solubility which can enhance its bioavailability (Starkloff et al. 2016).

6.6 3CLpro Inhibitors and Polymeric Nanoparticles

3CLpro, being the main SARS CoV2 protease enzyme of the viral genome processing machinery, is considered as a crucial therapeutic target for fighting against SARS CoV2. The National Health Commission, China, has approved lopinavir and ritonavir (HIV 1 protease inhibitors) as an off-the-cuff treatment drug to treat the SARS CoV2 infection. Hence, these two drugs are used as reference inhibitors. Other inhibitors of the active site of the 3CLpro include 10-hydroxyusambarensine, cryptoquindoline, 6-oxoisoiguesterin and 22-hydroxyhopan-3-one. The vicinity of the 3CLpro active cleft has highly conserved triad of serine residues (Ser139, Ser144 and Ser147) which has affinity for bifunctional aryl boronic acid compounds (Bacha et al. 2004). As the inhibition constant is very high for the boronic acid compounds, they can be used as antiviral compounds. In silico studies on systemic screening of 3CLpro inhibitors for antiviral activities against SARS CoV2 reveal the following lead candidates (Gyebi et al. 2020; Khan et al. 2020): alkaloids, terpenoids, paritaprevir and raltegravir. Among them, paritaprevir ($\Delta G = -9.8$ kcal/mol) and raltegravir ($\Delta G = -7.8$ kcal/mol) are highly effective against SARS CoV2. Peptidomimetic analogs, covalent inhibitors, small molecule inhibitors and non-covalent inhibitors of 3CLpro are also potent anticoronaviral agents (Wei et al. 2006; Ding et al. 2005; Jacobs et al. 2013; Pillaiyar et al. 2016).

G5 amine-terminated polyamidoamine (PAMAM) dendrimer is able to form conjugates with boronic acid compounds (Liu et al. 2019) and exert strong inhibitory activity on MERS CoV (Kandeel et al. 2020). Polymeric, metallic and lipid nanoparticles are appropriate nanocarriers for 3CLpro inhibitor molecules like paritaprevir (Abd Ellah et al. 2019). Raltegravir can be conjugated to gold nanoparticles for the targeted inhibition of viral 3CLpro (Garrido et al. 2015). Polymethyl methacrylate [PMMA] nanoparticles and ε-caprolactone-based nanoparticles (ECL) functionalized with specific targeting ligand can result in targeted delivery of rateglavir (Ogunwuyi et al. 2016). Antibody-functionalized nanocarriers, nanomaterials conjugated to receptor-specific peptides, nucleic acid sequences and ligands such as folates, vitamins or carbohydrates, PEGylated PMMA nanoparticles, gold nanoparticles and iron nanoparticles can enable site-specific delivery of small molecules which may be considered for the delivery of small 3CLpro inhibitor molecules (Klimkowska et al. 2019; Smith et al. 2015; Jahangirian et al. 2019). Recently discovered SARS CoV2 main protease inhibitors suitable for the treatment of COVID-19 are deltaviniferin, myricitrin, chrysanthemin, myritilin, taiwanhomoflavone A, lactucopicrin 15-oxalate, nympholide A, afzelin, biorobin, hesperidin and phyllaemblicin B. Delivery of 3CLpro inhibitors assisted by nanotechnology will be appropriate in the management of COVID-19 (Joshi et al. 2020).

6.7 Dendrimers, Metal Nanoparticles and Quantum Dots Can Deliver PLpro Inhibitors

SARS CoV2 PLpro has three major functions: proteolytic cleavage of viral polyprotein complex (refer Chap. 2), hydrolysis of cellular proteins, and ubiquitin and ubiquitin-like ISG15 (interferon-induced gene 15). Thus, PLpro behaves as a protease, a deubiquitinating (DUB) enzyme and also a deISGylating enzyme. These functions lead to escape of the host innate immunity by the virus. Chloroquine and formoterol are inhibitors of the active site domain of PLpro. Hence, these drugs can be potent inhibitors of SARS CoV2 infection (Ayra et al.). Chloroquine activity and the role of nanomaterials in enhancing its efficacy have already been illustrated in the earlier topics of this chapter. Formoterol is a bronchodialator which is used in treating breathing disorder induced by SARS CoV2 infection. Other PLpro inhibitors which can be repurposed into anti-SARS CoV2 drug include thiopurine compounds (6-mercaptopurine /6MP) and 6-thioguanine/6TG), tanishones, geranylated flavonoids, diarylheptanoid, naphthalene-based lead compounds, zinc and zinc conjugates (Chen et al. 2009; Park et al. 2012a, b; Cho et al. 2013; Han et al. 2005; Lindner et al. 2005).

Nanoencapsulation in functionalized carbohydrate polymers results in targeted and controlled release of formoterol in lung cells (Thi et al. 2009; Kaur et al. 2012; Kuzmov and Minko 2015). Magnetite nanoparticle surface engineered with polyvinyl alcohol (PVA) enables the sustained release and targeted delivery of 6-MP with lower therapeutic dose and lower side effects (Moustafa et al. 2018). Furthermore, CdSe/ZnS core–shell quantum dots are used as ideal carriers and also as a detection probe for the delivery of 6-TG and also for tracking the presence of thione group in the 6-TG structure (Grabowska-Jadach et al. 2020). High pay loads of 6-methylguanine can be entrapped in disulfide-based polyethylene glycol (PEG)-conjugated nanogels to achieve stimuli-responsive release property (Mokhtari et al. 2019). Thus, polymeric nanoparticles, magnetite nanoparticles and core–shell quantum dots can be used in the COVID-19 drug design and detection probe identification.

6.8 Nanoformulation and PLGA Nanoparticles for the Targeted Delivery of Dolutegravir and Bictegravir

$2'$-OMTase (nsp10) inhibitors are potential druggable candidates for SARS CoV2 infection. Alkaloids, antivirals, cardiac glycosides and steroids are capable of inhibiting the active site of 2′OMTase. These candidates can therefore be used for SARS CoV2 therapy (Decroly et al 2008). The lead druggable candidates targeting 2′OMTase are dolutegravir ($\Delta G = -94$ kcal/mol) and bictegravir ($\Delta G = -84$ kcal/mol) (Khan et al. 2020). Dolutegravir is an organofluorine monocarboxylic acid amide and an organic heterotricyclic compound. Dolutegravir and bictegravir inhibits the stability of RNA and its transition by effectively interfering with the mRNA capping activity of 2OMTase.

A nanoformulation containing poloxamer and dolutegravir with sustained release property is applicable for human use (Sillman et al. 2018). Poly(lactic-co-glycolic acid)-loaded nanoparticles formulated using an oil-in-water emulsion protocol can enable the sustained release of bictegravir achieving high effect at low concentrations (Mandal et al. 2019).

6.9   Nanocarriers for NTPASE/Helicase Inhibitors

SARS CoV NTPase/helicase inhibitors exhibit anti-MERS and anti-SARS activities. Aryl diketoacid is one such inhibitor capable of interfering with genome replication in coronaviruses (Lee et al. 2009). Bismuth salts such as bismuth potassium citrate (BPC) and ranitidine bismuth citrate (RBC) are inhibitors of ATPase activity and suppressors of helicase activity in SARS CoV2 nsp13.

Water-in-oil (w/o) emulsions of bismuth salt nanoparticles exhibit anti-SARS CoV activity (Chen et al. 2010) on par with the clinically used drug, colloidal bismuth subcitrate (CBS) (Shu et al. 2020), thus offering a new strategy for nanomedicine-based treatment for COVID-19. Stage 1 in the SARS CoV2 life cycle (refer Chap. 3) reveals clathrin-dependent acidification mediating entry of the virus into the host cell cytoplasm. Nanocapsules made of poly(ethylene glycol)-block-poly(lactide-co-glycolide) (PEG-PLGA) containing diphyllin strongly inhibit ATPase activity and prevent the endosomal acidification (Hu et al. 2017). This event has the potential to prevent the SARS CoV2 internalization and subsequent replication. Integrating the arydiketoacids, BPC and RBC with nanotechnology concepts will open a novel avenue for repurposing anti-SARS CoV2 therapeutics (Shu et al. 2020). Multifunctionalized nanoparticles (refer Chap. 4) hosting helicase inhibitors (the main cargo) and the nuclear localization signal (the targeting ligand) can lead to site-specific delivery without side (Kang et al. 2010). Similarly, clinically approved nanoparticles can also be appropriately functionalized to deliver the helicase inhibitors to get high therapeutic efficacy (Datta and Brosh 2018).

6.10   Potential Nanoparticles for the Delivery of RDRP Inhibitors

RdRp of SARS CoV2 shows sequence identity with the counterpart of other positive-sense RNA viruses. Hence, the RdRP inhibitors of positive-sense virus can be used to target SARS CoV2 genomic replication. Sofosbuvir is an inhibitor of SARS CoV2 RdRp with potential therapeutic effect for COVID-19. Antiviral molecules like remdesivir, galidesivir, ribavirin, favipiravir (Gao et al. 2020), filibuvir, cepharanthine, simeprevir, tegobuvir (Ruan et al. 2020) and aurin tricarboxylic acid (ATA) (Savarino et al. 2006) can easily target the active site of RdRp suggesting an opportunity for development of promising SARS CoV2 therapeutic.

Among the RdRp inhibitors, gold nanoparticle-bound ribavirin exhibits strong inhibitory action against many RNA viruses in vitro. The antiviral activity of nanoparticle-bound drug is higher than the free drug (Ahmed et al. 2018). ATA directly binds to the p38 mitogen-activated protein kinase (MAPK) and inhibits the enzyme activity both in vivo and in vitro (Tsi et al. 2002). However, a nanoformulation of ATA with biocompatible and biodegradable polymers inhibited MAPK with high efficacy, thus suggesting a valid approach for COVID-19 treatment (Loretz et al. 2006; O'driscoll et al. 2019). Nonetheless, as RdRp is hotspots of mutation in SARS CoV2 genome (refer geographical distribution in Chap. 1) there are lots of opportunities for drug resistance and loss of binding affinity to drug molecules (Pachetti et al. 2020). This is a sensitive challenge to be addressed while repurposing RdRp inhibitors for combating SARS CoV2 infection. Monodispersed quasi-spherical gold nanoparticles of 7 nm size capped with gallic acid were found to be virucidal by preventing the viral attachment to the host cell in herpes simplex virus model. These gold nanoparticles kill resistant strains also and function as polymerase inhibitor. As SARS CoV2 RdRp is predominant in gene expression and reported as the mutation hotspot, functionalized gold nanoparticles can be used to target the attachment and polymerase enzyme (Halder et al. 2018).

6.11 PLGA and Metal Nanoparticles Conjugated Serine Protease Inhibitors

TMPRSS2 is a membrane-anchored serine protease found in the epithelial cells of nasal lining, respiratory lining and lung alveoli (Antalis et al. 2011). TMPRSS2 has two major functions: (i) promoting viral entry and (ii) initiating pathological and immunological response in the host cell. TMPRSS2 cleaves the nucleophilic serine residues present in the S glycoprotein of SARS CoV2 and primes the spike to fuse with the host cell for infection. In SARS CoV2-infected alveolar epithelial cells, TMPRSS2 genes and genes involved in inflammatory response are also activated. As a result, priming of spike protein is always accompanied by inflammatory response (Sungnak et al. 2020). Thus, inhibiting TMPRSS2 serine protease by inhibitors or by antibodies present in the convalescent sera could be an effective treatment for COVID-19 (Uskokovic 2020).

FOY-305 or the camostat mesilate or camostat monomethane sulfonate (Fig. 6.1) is a synthetic serine protease inhibitor (Kitamura and Tomita 2012) with anti-inflammatory activity and capable of suppressing the expression of the cytokines interleukin-1beta (IL-1b) and interleukin-6 (IL-6), which brands it being investigated for SARS CoV2 therapeutic activity. Of TMPRSS2 serine protease (Hoffmann et al.

2020; Bittmann et al. 2020), oil-in-water nanoemulsion of poly(lactic-co-glycolic acid) (PLGA) can be used as carriers for the controlled and efficient release of camostat mesilate (Yin et al. 2006). Henceforth, nanoencapsulated serine protease inhibitor may offer the dual advantages of suppressing the viral attachment and inflammation response leading to an appropriate SARS CoV2 treatment regime. Gold nanoparticles functionalized electrostatically or covalently with the serine protease inhibitor, and the targeting antibodies exhibit targeted intracellular delivery of the cargo, offering effective activity (David et al. 2018). Serine protease inhibitors conjugated to zinc oxide nanoparticles and titania nanoparticles exhibit high storage stability and high affinity (low Ki value) compared to the free inhibitor. This stability and high affinity for the target enzyme will improve the applicability of the inhibitor–nanoparticle conjugate for medical purposes (Billinger et al. 2019). Hence, gold nanoparticles, zinc oxide nanoparticles and titania nanoparticles can also be used for the effective targeted inhibition of SARS CoV2 serine proteases.

6.12 Nanoparticles of Phospholipids and Carrageenan for Umifenovir Delivery

Umifenovir (also called Arbidol, Fig. 6.1) is an indole-derivative exhibiting direct antiviral action against a broad range of enveloped and non-enveloped viruses. Considering SARS CoV2, Arbidol was effective in vitro by interfering with S glycoprotein binding to ACE2 (Huang et al. 2020; Wang et al. 2020a, b). Arbidol is capable of preventing the cleavage of S glycoprotein and the consequent virus-ACE2 binding in silico. This is due to its interaction with critical aromatic aminoacid residues on the spike protein which are needed for host cell fusion (Vankadari 2020). In order to derive the complete benefits of Arbidol, it has to bypass the first-pass metabolism in the liver and reach the lungs (Chenthamara et al. 2019). This is enabled by targeted pulmonary delivery using nanocarriers including liposomes, polymeric nanocapsules, solid-lipid nanoparticles and dendrimers (Zhang et al. 2010). As Arbidol can be optimally encapsulated in carrageenan (Aleksandrovich et al. 2016), the targeted delivery of Arbidol can be used as a treatment approach for SARS CoV2 infection. Pharmaceutical composition containing Arbidol in the form of phospholipid nanoparticles was patented (Archakov et al. 2012). This product can emerge as an efficient ACE2 inhibitor to combat SARS CoV2 infection.

6.13 Targeting the Inflammation and Cytokine Storm

SARS CoV2-infected lungs show considerable degree of inflammation resulting in alveolar damage and lesions during early as well as severe stages. These features are due to viral multiplication and host immune reaction (Wu et al. 2020) suggesting a therapeutic strategy involving amelioration of inflammation. Protein cage nanoparticles are capable of activating iBALT (inducible bronchus-associated lymphoid tissue) which leads to the alleviation of the lung inflammation and the viral titer (Wiley et al. 2009). Next, the anti-inflammatory action of leukemia inhibitory factor (LIF) can be availed for inhibiting the inflammation. LIF, if administered intranasally or intravenously, can prevent alveolar epithelial damage and vascular leakage and promote regeneration of lung alveoli (Metcalfe 2020; Quinton et al. 2012). PLGA nanoparticles can enhance the targeted delivery of LIF (Metcalfe et al. 2015) which can be efficient in combating SARS CoV2-induced inflammation (Uskokovic 2020).

Cytokine storm or hypercytokinemia is a prominent feature of SARS CoV2 infection (Mehta et al. 2020), which can be targeted by immunomodulators, antioxidants and gene silencing therapy using siRNA. Lipid nanoparticles made out of squalene after being functionalized with adenosine and alpha tocopherol possess anticytokine storm activity which leads to the down-regulation of inflammation. Adenosine functions as immunomodulator, while tocopherol functions as antioxidants in reducing the oxidative damage due to inflammation. This concept can be used to treat inflammation associated with COVID-19 (Dormont et al. 2020).

SiSC2 and siSC5 are the potent siRNA candidates with prophylactic and therapeutic effect against coronavirus infection, and are capable of reducing the alveolar damage (Li et al. 2005). But, effective intranasal delivery of siRNA using nanostructures will be useful for ameliorating the lung damage without off-target side effects (Li et al. 2005). Nanoparticles, cationic lipid structures, antibodies and aptamers (refer Chap. 5) are acceptable delivery strategies for safer and site-specific delivery of siRNAs (Chen et al. 2020).

SARS CoV was reported to directly upregulate p38 activity and induce inflammation. On this basis, p38 inhibition was used as an anti-SARS CoV therapeutic strategy in mouse model. Owing to homology with SARS CoV, SARS CoV2 binds to ACE2 receptor for host entry and inactivates the latter, allowing angiotensin 2-mediated activation of p38 (Grimes and Grimes 2020). Hence, P38 inhibitors can be considered for fighting COVID-19 infection also (Mizutani 2007). Nanoformulation of p38 MAPK inhibitors may enable controlled release and targeted delivery. Reported nanoformulations of the p38 MAPK inhibitors are GSK 678361A encapsulated in poly(lactic-co-glycolic acid) nanoparticles and PH797804 polymeric nanocrystals (Maudens et al. 2018; Bains et al. 2016). We imply that p38 inhibitors can be delivered using nanocarriers for better anti-COVID-19 action.

6.14 Viroporin Inhibitors

Viroporin is a membrane protein which functions as a proton channel. It is responsible for maintaining a low pH in the endosomes which is suitable for dissolving the nucleocapsid and releasing the viral genome into the host cytoplasm (Schoeman and Fielding 2019). Inhibitors of viroporin can therefore be an ideal therapeutic tool for targeting SARS CoV2 genome entry. Procoxacin (also called salinomycin) is an inhibitor of viroporin protein, and it exhibits antiviral activity against respiratory virus (Jang et al. 2018). Thus, salinomycin can emerge as a drug for targeting SARS CoV2 genome entry. Lipid nanostructure-mediated delivery of salinomycin may bypass the "first-pass" metabolism, resulting in targeted release, enhanced bioavailability at the intrapulmonary site with low side effects (Pindiprolu et al. 2020).

6.15   Nanostructures for Targeting SARS CoV2 Co-Receptors

Coronaviruses are capable of using secondary receptors or co-receptors along with the primary receptors for gaining access into the host cell. For example, MERS CoV uses sialic acid chains as a co-receptor in addition to its main receptor, DPP4. Similarly, SARS CoV depends on heparan sulfate proteoglycan chains as co-receptor in addition to ACE2 (Li et al. 2017; Milewska et al. 2014; Lang et al. 2011). The spike protein of SARS CoV2 has high binding affinity for the specific motif of heparan sulfates containing repeated units of IdoA2S. Nanoparticles are good at targeting heparan sulfates. Dendritic polyglycerol sulfate nanogels imitate cellular heparan sulfate and shield the viral surface from entering into host cell, thus preventing infection (Dey et al. 2018). Tellurium nanostars functionalized with bovine serum albumin have the ability to bind to the heparan sulfate analogs and interfere with the viral access into the host cell. Hence, dendritic polyglycerol and tellurium nanostars could be used for targeting the co-receptor-mediated SARS CoV2 entry into the host cell (Zhou et al. 2020).

Dipeptidyl peptidase 4 (DPP4) or the cluster of differentiation 26 (CD26) is a serine exopeptidase expressed as surface receptors in many tissues including lungs. Though ACE2 is a primary receptor for SARS CoV2, DPP4 has recently been reported as a new secondary receptor target for SARS CoV2 spike protein (Vankadari and Wilce 2020; Iacobelis 2020; Lu et al. 2013). Therefore, DPP4 modulation or inhibition may prevent infection and may evolve as a promising strategy in the management of COVID-19. Sitagliptin is an inhibitor of DPP4 (Xiao et al. 2014). Poly(lactic-co-glycolic acid) (PLGA) nanoparticles synthesized by nanoprecipitation cum solvent evaporation method show high entrapment efficiency and controlled release pattern for sitagliptin. The encapsulated nanosystem is capable of being engineered with antibody as the targeting ligand (Thondawada et al. 2018). Hence, sitagliptin-loaded PLGA nanoparticles functionalized with DPPR-targeting antibodies can be a promising treatment mode for SARS CoV2 infection.

6.16   Lipid Raft Inhibitors

Lipid rafts are specific liquid-ordered microdomains of plasma membrane containing high quantities of sphingolipid and cholesterol (Simons and Toomre 2000). Lipid rafts accommodate signaling molecules and transport proteins responsible for the pathogenesis of many diseases and also in eliciting immune response (Manes et al. 2003). Lipid raft is also an important motif in the pathogenesis of virus as it enables virus entry through endocytosis (van der Goot and Harder 2001). Coronaviruses subvert raft-associated signaling and utilize lipid raft for entry and cell fusion which enables their efficient replication while blocking the immune response that is elicited by the target cells (Choi et al. 2005). Elucidation of the mechanism of hijack of host raft domains by SARS CoV2 will afford new therapeutic insights for the prevention and/or treatment of COVID-19. Lipid raft studies also help to understand the molecular pathogenesis behind enhanced cytokine storm. Cyclodextrin is important lipid raft inhibitors which have the ability to prevent lipid-draft-dependent cellular entry of coronavirus including SARS CoV2 (Wang and Hajishengallis 2008; Baglivo et al. 2020).

Cyclodextrins are starch-based glucopyranose polymers with truncated cone-shaped hydrophobic cavity (Lembo et al. 2018). The nanotized form of cyclodextrin consists of solid nanoparticles with cross-linked cavities. These structures are called as nanosponges (Swaminathan et al. 2016). Cyclodextrin nanosponges can be a promising platform for investigating advanced antiviral therapeutic treatments to fight viral infections including COVID-19.

6.17   SKP2 Inhibitors

Autophagy is a normal catabolic process involving the destruction of cells by protein degradation. Autophagy is scaled up during viral infection in order to destroy the viruses and regulate the inflammatory responses. (On the other hand, viruses exploit the autophagy for the replication and release phases and also for the immune evasion.) S phase kinase-associated protein 2 (SKP2) tags beclin-1 and executes lysine-48-linked poly-ubiquitination and proteasomal degradation of beclin-1. Small molecule inhibitors of SKP2 can increase the beclin level to enhance autophagy. This mechanism was involved in interfering with the replication of MERS CoV genome. This confirms the prospect of SKP2 inhibitors for SARS CoV2 therapy (Gassen et al. 2019). Niclosamide (chlorinated salicylanilide) is an anthelminthic drug possessing broad-spectrum antiviral effect. This drug is also effective against SARS CoV and MERS CoV. Niclosamide is an inhibitor of SKP2. However, a pharmacokinetic demerit of low absorption warrants design and development of drug formulation for enhanced solubility, absorption, targeted effect and sustained release of the drug (Yang and de Villiers 2005). Undoubtedly, nanocarriers can circumvent this flaw. In this context, chitosan nanoparticles may be of immense use. Water dispersible chitosan nanoformulations of niclosamide can be obtained by glutaraldehyde cross-linking via covalent chemical interaction between positively charged amino groups of chitosan and aldehyde group of glutaraldehyde (Naqvi et al. 2017). In addition, a formulation of niclosamide and solid-lipid nanoparticles (SLNs) is an ideal carrier system for effective intracellular delivery of niclosamide (Pindiprolu et al. 2019).

6.18 STAT3 Inhibitors

Signal transducer and activator of transcription-3 (STAT3) is a key regulator of host immune and inflammatory responses. STAT3 has both proviral and antiviral activities, and it is involved in the regulation of viral replication and pathogenesis thereafter. Here, we consider the proviral activity to use STAT3 inhibitors as an effective anti-SARS CoV2 target. STAT3 is constitutively phosphorylated at tyrosine (Tyr)-705 and slightly phosphorylated at serine (Ser)-727. SARS CoV infection leads to p38 MAPK pathway-induced STAT3 dephosphorylation at Tyr-705 in the host cell, in order to promote viral replication (Mizutani et al. 2004a, b). STAT3 modulators can prevent viral replication and can be an efficient anti-SARS CoV target. Niclosamide and silibinin (a favolignan from milk) are reported STAT3 modulators. Niclosamide inhibits STAT3 and prevents viral replication. Silibinin can target both the host cytokine storm and the virus replication as well to clinically manage COVID-19/SARS CoV2 infection (Bosch-Barrera et al. 2020). Poor bioavailability is however the disadvantage of silibinin and niclosamide. Hence, nanoparticles-based formulations and carrier system are recommended for enhancing the bioavailability for wide spectrum applications (Milad et al. 2020).

6.19 Lectin Inhibitors

As described in stage 1 of SARS CoV2 life cycle (Chap. 3), lectins are used as ACE2-independent binding cum infection pathway. Hence, targeting the host cell lectin can prevent virus entry. SARS CoV2 infectivity can be blocked by those lectins that are specific for the glycans present in the spike glycoprotein. As galactose sugar moieties can bind with lectins, galactose-modified nanomicelles can be used to target the ACE-2 independent infection pathway (Chiodo et al. 2020; Takada et al. 2004). Mannose-binding lectins are also efficient in binding to SARS CoV spike proteins glycan residues (Keyaerts et al. 2007).

6.20 Nanosponges

Nanosponges are nanoparticles cloaked in capsules or vesicles formed out of membranes from human lung cells and immune cells. They act as decoys to counteract SARS CoV2 and prevent host cell infection. Nanosponges, as the name indicates, are capable of soaking the viruses, and they show a dose-dependent decrease in SARS CoV2 infectivity. Nanosponges can hence decrease the capacity of the virus to pass into the host cell for exploiting its resources, to replicate and to produce additional infectious viral particles (Xinyi 2020; Zhang et al. 2020).

The potential nanotechnology-based strategies for advanced SARS CoV2 therapy discussed so far are summarized in Table 6.2.

6.21 Translational Nanomedicine Avenues for SARS CoV2 Infection

Many viruses depend on intracellular iron for their propagation. Thus, iron supply is obligatory for the replication of many viruses including CoVs. Thus, translational research involving manipulation of iron metabolism and associated disorders like hemoglobin dysfunction and hypoxic states will be a valid management strategy for COVID-19. Iron chelation, administration of heme metabolic regulators like vitamins, and antioxidants like glutathione are choices of iron metabolism manipulators and ferroptosis regulators (Liu et al. 2020a, b). Concepts used in translational medicine for SARS CoV2 management (Cavezzi et al. 2020) are shown in Table 6.3, and the potential drug compounds are illustrated in Fig. 6.2 and explained under this topic.

SARS CoV2 mimics a molecule called hepcidin (an inhibitor of the ferroportin-based cellular iron export). In fact, the C terminus of the spike glycoprotein tail is similar to the hepcidin. Thus, tissue ferritin overload is one of the features of SARS CoV2 infection, which is a negative prognostic factor. Antihepcidin molecules or small molecule inhibitors of hepcidin (hepcidin antagonists) or ferroportin agonists may block the SARS CoV2's hepcidin-like motif, promote export of iron from lungs, ameliorate the ferritin overload and hence open a new therapeutic avenue (Shah et al. 2020; Liu et al. 2020a, b; Yilmaz and Eren 2020).

Tocilizumab is a monoclonal antibody raised against hepcidin (i.e., tocilizumab is an hepcidin antagonist). It is possible to coat tocilizumab on to gold nanoparticles by physical modification creating a hybrid with high stability. Tocilizumab–gold nanoparticle conjugate could emerge as a therapeutic tool as well as a diagnostic approach involving molecular recognition of monoclonal antibodies (Lee et al. 2014).

TMPRSS6 gene encodes a serine protease TMPRSS6. This enzyme regulates hepcidin gene expression. Lipid nanoparticle-encapsulated siRNA targeted to TMPRSS6 gene can serve as an upstream regulator of hepcidin (Schmidt and Fleming

Table 6.2 Promising nanostrategies toward COVID-19 therapy

Drugs	Potential nanostrategies	Possible advantages	References
Nanobodies Antibodies	Liposomes Polyethylene glycol-poly(ε-caprolactone) nanocapsules PLGA-PEG nanoparticles	Targeted delivery Lesser degradation of antibodies High antigen recognition ability High paratope availability	Abraham (2020) Wrapp et al. (2020)
Chloroquine	Formulation of PLA nanospheres Chitosan tripolyphosphate-conjugated nanochloroquine Gold–chloroquine nanoconjugate Gelatin nanocarriers	High biodegradability Controlled release Stimuli-responsive drug release High encapsulation efficiency and sustained release	McKee et al. (2020)
SiRNA	Liposomes Polyaspartamide Polyethyleneimine-modified magnetic Fe_3O_4 nanoparticles Carbon apatite nanoparticles	Targeted therapy Treatment of breathing issues Targeted delivery and contrast-enhanced imaging PH-triggered release	Ghosh et al. (2020)
Ivermectin	Liposomes SLNs Nanosuspensions	High encapsulation efficiency High solubility High biocompatibility Controlled release Lesser drug resistance Lesser side effects	Ahmadpour et al. (2019) Heidary and Gharebaghi (2020)
Paritaprevir	PAMAM dendrimer Lipid nanoparticle	Targeted inhibition of 3CLpro	Khan et al. (2020)
Raltegravir	ε-Caprolactone Gold nanoparticle	Targeted inhibition of 3CLpro	Garrido (2015)
6- Mercaptopurine 6-Thioguanine 6-Methylguanine	Magnetite nanoparticles CdSe/ZnS core–shell quantum dots Nanogels	Targeted inhibition of PLpro	Mokhtari et al. (2019); Arya et al. (2020)
Bictegravir Dolutegravir	Poloxamer nanoformulation PLGA nanoparticles	Targeted inhibition of 2 OMTase	Mandal et al. (2019) Khan et al. (2020)
Bismuth salts	Bismuth salt nanoparticles	Targeted inhibition of NTPase/helicase	Shu et al. (2020)

(continued)

Table 6.2 (continued)

Drugs	Potential nanostrategies	Possible advantages	References
Diphyllin	PEG-PLGA nanoparticles	Effective inhibition of endosomal acidification	Hu et al. (2017)
Ribavirin and aurin tricarboxylic acid Gallic acid	Gold nanoparticles Polymeric nanoformulation Quasi-spherical gold nanoparticles	Strong inhibition of RdRp Evades drug resistance	Ahmed et al. (2018) Savarino et al. (2006) Halder et al. (2018)
Serine protease inhibitor	Zinc and titania nanoparticles	High affinity for serine protease High stability	David et al. (2018) Billinger et al. (2019)
Umifenovir	Phospholipid nanoparticle Carrageenan nanocapsule	Targeted and controlled release	Aleksandrovich et al. (2016)
Anti-inflammatory agents	PLGA nanoparticles Lipid nanoparticles	Reduces alveolar lung damage Decreases oxidative damage by immunomodulation	Thondawada et al. (2018) Dogaru et al. (2020)
Procoxacin	Lipid nanostructures	High bioavailability, intrapulmonary targeting, low side effects	Pindiprolu et al. (2020)
Heparan sulfate mimics	Dendritic polyglycerol	Prevents viral binding to co-receptors	Zhou et al. (2020)
Sitagliptin	PLGA nanoparticles	Encapsulation efficiency and controlled release	Thondawada et al. (2018)
Niclosamide	Chitosan nanoparticles	Enhances solubility and absorption	Naqvi et al. (2017)

2014). Thus, lipid nanoparticle-encapsulated hepcidin may be a hopeful translational therapeutic strategy to treat iron overload in SARS CoV2 infection.

Ferroportin agonists can be used to treat hemoglobin dysregulation. Erythropoietin is a ferroportin agonist, which is known to influence the ferroportin mRNA expression in a dose-dependent manner (Srai et al. 2010). Further, poly(DL-lactide-co-glycolide) nanoparticles–erythropoietin conjugate prepared using double emulsion method (w/o/w) was found to be released in a controllable fashion with high stability than the free erythropoietin (Fayed et al. 2012).

SARS CoV2 gains entry through CD147, CD26 and other receptors located on erythrocyte or blood cell precursors and binds with the hemoglobin. This leads to hemoglobin denaturation and release of free heme causing dysfunction of hemoglobin. Due to dysregulated hemoglobin, the metabolism shifts the ferrous–ferric equilibrium toward the active ferric form. In this form, the oxygen binding

Table 6.3 Possible avenues for translational nanomedicine to combat SARS CoV2 infection

Concepts in SARS CoV2 infection	Avenues for COVID-19 translational medicine	Nanoparticles	References
Ferritin overload in tissues Dysfunction of ferroportin (exporter protein)	Hepcidin antagonists (e.g., tocilizumab) Ferroportin agonists (thiazolidinone derivatives)	Gold nanoparticles Lipid nanocapsules	Lee et al. (2014) Srai et al. (2010) Fayed et al. (2012)
Hemoglobin dysfunction	Erythropoietin administration Blood transfusion Glucose/heme—arginine infusion Administration of repurposed drugs capable of targeting CD147/CD26 receptors on erythrocytes	PLGA nanocapsules	Cavezzi et al. (2020)
Reduction in nitric oxide	Ascorbic acid supplement	Polymeric nanogels Chitosan nanocapsules	Othman et al. (2018)
Ferroptosis	Ascorbic acid and reduced glutathione	Gold nanoanchors	Luo et al. (2016)
Unregulated RNA polymerase activity Viral invasion and virulence	Zinc administration owing to antiviral, RNA polymerase inhibition, prevention of viral entry and virulence	Zinc oxide nanoparticles PEGylated zinc oxide nanoparticles	Ghaffari et al. (2019) Tavakoli et al. (2018)
Altered hemoglobin level Down-regulation of hepcidin antagonist pathway	Vitamin D3 (cholecalciferol) administration to restore hemoglobin level and activate hepcidin antagonist pathway	Oil–water nanoemulsion Liposomes Solid-lipid nanoparticles Polymeric nanocapsules	Ozturk et al. (2015) Walia et al. (2017), Knudsen et al. (2012), Sonawane et al. (2014) Ramezanli et al. (2017) Lalloz et al. (2018)
Hemoglobin denaturation Oxidative damage to ferroportin receptor	Melatonin decreases heme protein denaturation, prevents ferroptosis and neutralizes hypoxia	Zein nanoparticles Solid-lipid nanoparticles	Li and Zhao (2017)
Altered red cell homeostasis and hemoglobin dysfunction	Polyphenols like curcumin normalize red cell homeostasis and hemoglobin saturation and transport	Curcumin analog nanoparticles	Francis et al. (2015) Francis et al. (2018) Devasena and Rajasekar (2014)

(continued)

Table 6.3 (continued)

Concepts in SARS CoV2 infection	Avenues for COVID-19 translational medicine	Nanoparticles	References
Hypoxia-induced, fibrinolytic-resistant parafibrin formation associated with altered coagulation cascade	Anticoagulant heparin prevents fibrin formation and the coagulation cascade	Heparin nanoassemblies	Bellido et al. (2015), Salmaso and Caliceti (2013) Socha et al. (2009)
Mitochondrial depolarization and degeneration	Curcumin, anthocyanins Ascorbic acid Ubiquinol Nicotinamide mononucleotide Specific mitochondria-targeted antioxidants	Nanocurcumin Nanocurcumin analog	Priya et al. (2020) Francis et al. (2015) Francis et al. (2018) Devasena and Rajasekar (2014)

capacity (i.e., the normal hemoglobin function is decreased) leads to low-oxygen symptoms like hypoxia and hypoxemia. Therefore, SARS CoV2 is also characterized by hypoxemia and systemic hypoxia. Nitric oxide (NO) secretion usually compensates hypoxia by inducing vasodilation, but circulating cell-free hemoglobin would impair NO bioavailability and induces acute respiratory distress syndrome. Further, the low hemoglobin concentration and low hemoglobin saturation in SARS CoV2 infectious stage lead to vasoconstriction and pulmonary hypertension. These two critical conditions may be targeted for translational medicine development. High oxygen supply at low pressure, prostanoids, acetazolamide, calcium channel blockers, sildenafil and nitroglycerine-based drugs may be potential compounds to target the vasoconstriction.

It is known that SARS CoV2 uses ACE2 as an entry gate to host cell. But, ACE2 is capable of increasing the synthesis of the vasodilator angiotensin 1–7, post-infection. The possibility of functionalizing ACE2 on to the surface of quantum dots and nanoflowers has already been proposed (Aydemir and Ulusu 2020). Thus, the hypoxia-induced vasoconstriction can be managed. Iron chelators like deferiprone or deferoxamine may help in neutralizing overloaded cellular iron content.

Ascorbic acid is capable of doing two crucial roles: (i) maintaining the iron in ferrous state in order to enable binding with hemoglobin. This leads to good oxygen supply (ii) restoring nitric oxide (NO) and improving the heme-cell balance. Nanotechnology formulation can promote stability and therapeutic activity of ascorbate molecules. An optimized nanogel formulation containing hydroxypropyl methyl cellulose gel helps in the sustained release of ascorbate (Duarah et al. 2017). In addition, ascorbic acid encapsulated into chitosan nanoparticles by ion gelation method can also lead to effective pharmacological effect (Othman et al. 2018).

Ferroptosis is the iron-dependent cell death caused by ferritin overload. Ferroptosis induces oxidative stress and antioxidant imbalance. Reduced glutathione (GSH)

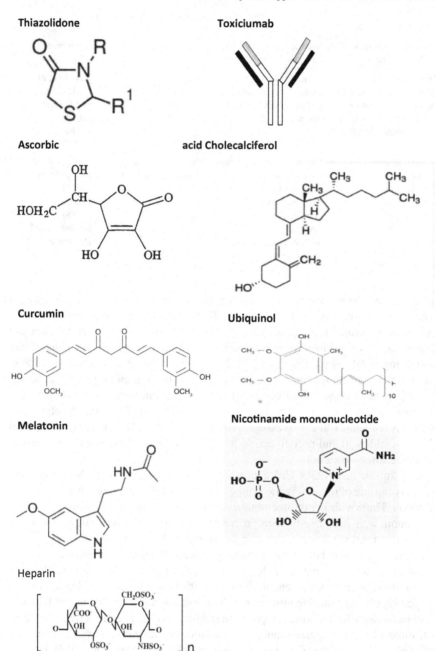

Fig. 6.2 Promising translational medicines to combat SARS CoV2 infection

can counterbalance the post-ferroptosis effect. GSH can also normalize oxidation state of iron toward ferrous state in conjunction with ascorbic acid. The thiol group of GSH is crucial for the antioxidant activity, and hence it is essential to preserve this moiety for better efficacy. Gold nanoanchors are used for capturing the GSH and gain high pharmacological activity when compared to free GSH. Gold nanoparticles capped with dihydrolipoic acid (AuNP-DHLA) can bind to GSH in the presence of N-ethyl-N'-(3-dimethylaminopropyl)carbodiimide (EDC) and N-hydroxysuccinimide (NHS) as a coupling agent. The carboxyl moiety of the DHLA bonds with the amine group of the GSH to generate a nanocomplex with high antioxidant activity (Luo et al. 2016).

Zinc is considered as a key micronutrient in translational medicine for SARS CoV2 due to the following activities favorable to the host cells:

- Anti-RNA virus activity
- Effect on RNA polymerases
- Effect on viral entry
- Nullifying the effect on virulence factors of the virus
- Enhancing the hemoglobin–oxygen affinity and neutralizing the red cell or hemoglobin.

Zinc oxide nanoparticles with and without PEG chains are capable of exhibiting high antiviral efficacy (Ghaffari et al. 2019; Tavakoli et al. 2018).

Vitamin D3 or cholecalciferol regulates hepcidin–ferroportin axis by activating hepcidin antagonist pathway and restoring the hemoglobin levels. Studies have insisted the occurrence and seriousness of vitamin D3 deficiency due to a lack of sun exposure, extensive use of sunscreens and consequent block in vitamin D synthesis and poor dietary intake. Further, the high phytate and fiber content of vegetarian diets may also reduce vitamin D3 absorption. All these problems alarm the scientific and research society to develop a nanotechnology-based delivery system for getting sufficient quantity of vitamin D3 from food and beverage fortification. Oil–water nanoemulsions (Ozturk et al. 2015; Walia et al. 2017), liposomes (Knudsen et al. 2012), solid-lipid nanoparticles (Sarangi et al. 2018; Sonawane et al. 2014) and polymeric nanocapsules (Ramezanli et al. 2017; Lalloz et al. 2018) are highly efficient in the delivery of functional vitamin D3.

Melatonin helps in preventing the (i) denaturation of hemoglobin, (ii) oxidative stress-induced damage of ferroportin (i.e., prevents ferroptosis) and (iii) hypoxia. Melatonin loaded on to zein nanoparticles exhibits different morphologies including: nanorods, nanofilaments and nanoglobules. The nanosystem displayed a near-zero-order release (Li et al. 2017) suggesting a potential prolonged and controlled release system for SARS CoV2 management. Eudragit S100 polymeric nanocapsules and nanospheres encapsulating melatonin prepared by nanoprecipitation method are capable of showing higher activity (Schaffazick et al. 2005).

Curcumin, a polyphenol from the rhizome of *Curcuma longa,* functions as heme-oxygenase activator. This may be a potential inhibitor of hemoglobin denaturation and dysfunction. Curcumin and its derivatives, especially in the form of nanocurcumin or in the form of chitosan nanoformulation, show better activity than the native compound (Francis et al. 2015). Nanotized curcumin analog was able to show better effects in the lung cells exposed to inhalation toxicity (Francis et al. 2018). Nanocurcumin is also better than free curcumin in exerting the antioxidant effect (Devasena and Rajasekar 2014). Potential role of curcumin and its derivatives are discussed under a separate topic in this chapter.

The dysregulation of iron metabolism, ferroportin oxidative damage and hypoxia are capable of generating fibrinolysis-resistant parafibrin formation. Thus, impairment in the coagulation cascade is one of the characteristic features of SARS CoV2 infection, which claims anticoagulant administration. Heparin nanoassemblies obtained from O-palmitoyl-heparin and α-cyclodextrin in water are useful in antiviral applications including protection against respiratory syncytial virus (Bellido et al. 2015; Salmaso and Caliceti 2013; Socha et al. 2009). The antiviral activity depends on the sulfation degree of heparin (Fig. 6.2). Therefore, nanoformulations of heparin may be one of the choices to be exploited as a potential therapeutic approach to combat COVID-19. In addition to dysregulation in hemoglobin and iron metabolism, mitochondrial depolarization and degeneration are also a possible consequence of SARS CoV2 infection. SAR CoV2 depends on the mitochondrial degradation and dysfunction to induce apoptosis (Priya et al. 2020) as indicated by enhanced lactate-level post-infection. Hence, drugs targeted to the mitochondrial restoration may be an adjuvant therapy for managing COVID-19. Curcumin, anthocyanins, ascorbic acid, ubiquinol, nicotinamide mononucleotide and specific mitochondria-targeted antioxidants may be useful in this approach. Nanotechnology-based delivery system for curcumin and ascorbic acid has already been discussed under this section.

6.22 Curcumin: The Indian Solid Gold

The Indian traditional-based medicine offers least side effects over synthetic drugs. Curcumin ($C_{21}H_{20}O_6$; molecular weight 368.38 g/mol) or diferuloylmethane chemically called 1,7-bis(4-hydroxy-3-methoxyphenyl)-1,6-heptadiene-3,5-dione (Fig. 6.2) is a natural yellow compound derived from turmeric (*Curcuma longa*). It is widely used as a spice and coloring agent in food. The pleiotropic effects of curcumin include antioxidant, antitoxic, anticarcinogenic, anti-inflammatory and antimicrobial effects with excellent safety profile.

Curcumin has antiviral effects on virulent viruses such as human immunodeficiency virus (HIV), human T-lymphotropic virus type 1 (HTLV-1), hepatitis B virus (HBV), hepatitis C virus (HCV), human papilloma virus (HPV), herpes simplex virus (HSV), Japanese encephalitis virus (JEV), influenza and coxsackievirus (Karbalaei and Keikha 2019). Curcumin interferes with many different target points in the virus and the host cells to exert its antiviral effects.

The following reported properties of curcumin and its analogs (Karbalaei and Keikha 2019; Prasad and Tyagi 2015) can be utilized for targeting SARS CoV2.

- Curcumin and curcumin analogs inhibit infection and replication of viruses

- Inhibition of proteases
- Enhancing the effect of conventional therapeutic drugs and minimizing their side effects
- Increasing the anti-inflammatory signaling
- Blocking the pro-inflammatory signaling
- Curcumin can influence various immune responses such as mitogen-activated protein kinases (MAPKs), phosphoinositide 3-kinase/protein kinase B (PI3K/PKB), nuclear factor kappa B (NFK-B) pathways, as well as dysregulate the ubiquitin–proteasome system.

6.22.1 Curcumin Inhibits the Binding of Enveloped Virus to the Host Cell

Curcumin due to its hydrophobicity is capable of intercalating with the hydrogen bonds of lipid bilayer via its phenolic moiety. This leads to alterations in the membrane morphology, permeability and fluidity (Chen et al. 2013), thereby preventing the viral binding and fusion with the host cell (Mounce et al. 2017). Curcumin is expected to interfere with binding of the SARS CoV2 envelope to the lung cells and emerge as a treatment for COVID-19.

6.22.2 Curcumin Blocks the Binding Between S Protein and ACE2 Receptor

Curcumin binds to the S protein of SARS CoV2 as well as to the ACE2 receptor of the host cell with binding energy of −7.9 kcal/mol and −7.8 kcal/mol, respectively (Atala et al. 2020). Therefore, curcumin can target viral entry into the host cell and may be an appropriate therapeutic tool for COVID-19.

6.22.3 Curcumin Binds to Nucleocapsid Protein and Nsp10

Curcumin possesses strong binding affinity for nucleocapsid phosphoprotein and nsp10 as compared to chemical antiviral drugs such as ivermectin, azithromycin and remdesivir. The two-terminal phenolic moieties linked by the central 7-carbon chain allow for a symmetry which functions as a clamp in the active site of the SARS CoV2 proteins. The zinc finger groove in nsp10 protein binds to the transcription regulatory proteins which help in recognizing RNA and also in protein–protein recognition. Curcumin hijacks nsp10 by binding to the zinc finger groove (Suravajhala et al. 2020).

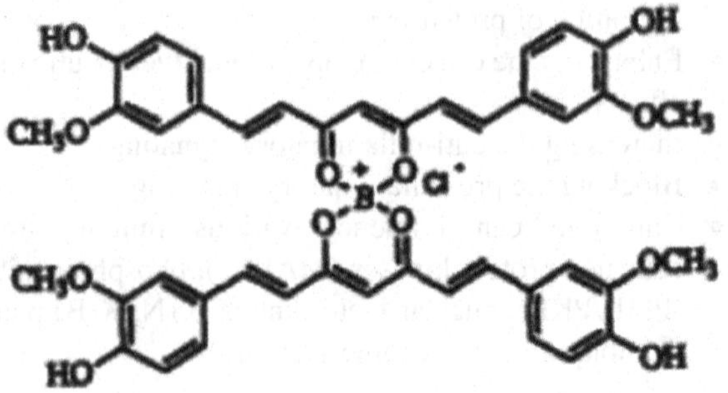

Fig. 6.3 Boron complex of curcumin

6.22.4 Boron Complex of Curcumin is Protease Inhibitors

HIV-1 protease inhibitors are reported to be SARS CoV main protease inhibitors. Later, it was also known that HIV-1 protease inhibitors such as lopinavir, ritonavir and saquinavir establish strong bonding with SARS CoV2 main protease also (Ortega et al. 2020). Boron complex of curcumin (formed by bridging of boron between carbonyl groups of two curcumin molecules; Fig. 6.3) is an inhibitor of viral proteases. But, the minimum inhibitory concentration is lower for boron complex when compared to free curcumin and this is because of high affinity of the complex to the active site cleft of the protease (Sui et al. 1993).

Curcumin inhibits replication of enveloped viruses including the SARS CoV by binding to the 3CLpro, the enzyme needed for the replication of SARS CoV genome (Mounce et al. 2017; Wen et al. 2007). Taken together, curcumin and boron complex of curcumin may be used as a therapeutic approach to target the proteases of SARS CoV2 and be used for managing COVID-19 in the future.

6.22.5 Curcumin Suppresses Cytokine Storm

The excessive pulmonary inflammatory responses or the "cytokine storms" are a prominent feature of SARS CoV2 infection. If unattended, the cytokine storm will be leading to pulmonary edema, atelectasis and acute lung injury (ALI) or fatal acute respiratory distress syndrome (ARDS). Curcumin impairs cytokine storm by regulating pro-inflammatory and anti-inflammatory factors such as IL-6, IL-8, IL-10 and COX-2, promoting the apoptosis of PMN cells and scavenging the reactive oxygen species (ROS) (Liu and Ying 2020; Liu et al. 2020a, b). 4-hydroxyphenyl unit is responsible for the anti-inflammatory activity which can be further amplified by the conjugation of acyl or alkyl or methoxy groups to the terminal benzene ring of the curcumin (Aggarwal et al. 2013). Thus, curcumin can be used to treat SARS CoV2 infection. The anticytokine storm activity of curcumin is mainly attributed to the following strategies:

(A) *Inhibiting pro-inflammatory cytokine pathway*
Inhibition of pro-inflammatory cytokine pathway is mainly due to the inactivation of NFKB signaling pathway by the following mechanisms:

- Inhibiting the activation of $I\kappa K\beta$ (I kappa B kinase β) which downregulates the expression of inflammatory cytokines IL-8, TNF-α and IFN-γ
- Upregulating and stabilizing the $I\kappa BK\alpha$ and inhibiting its degradation thereby blocking the pro-inflammatory gene expression
- Activating AMPK (AMP-activated protein kinase)
- Inhibiting the cyclooxygenase 2 (COX-2) activity.

(B) Activating anti-inflammatory cytokine pathway
Curcumin increases the expression of IL10. IL10 decreases inflammation by reducing the release of TNF-α, IL-6 and ROS by inflammatory monocytes. IL10 downregulates the expression of intercellular adhesion molecule-1 (ICAM-1) and TNF-α levels in the lungs, both leading to a decrease in the myeloperoxidase-mediated lung damage.

(C) Upregulating antioxidant enzymes
Curcumin neutralizes the oxidative stress by acting as an antioxidant which is attributed to its carbonyl, methoxy, hydroxyl and beta diketo moieties, all of which possess the ability to scavenge free radicals (Vajragupta et al. 2003). Further, curcumin upregulates the activities of antioxidant enzymes like superoxide dismutase, glutathione peroxidase and catalase, all responsible for the antioxidant and reactive oxygen species scavenging effects (Devasena et al. 2002).

(D) Curcumin alters the mitochondrial membrane potential

Virus infection affects mitochondrial dynamics. Altered mitochondrial dynamics enhances viral pathogenesis. As SARS CoV2 infection leads to mitochondrial depolarization and degeneration, drugs targeted to mitochondrial restoration may be an adjuvant therapy for managing COVID-19. Curcumin is useful in this approach as it is capable of restoring the mitochondrial dynamics and potential (Chen et al. 2013; Jaruga et al. 1998).

6.22.6 Nanotechnological Intervention of Native Curcumin

Regardless of its multiple benefits, some disadvantages such as high hydrophobicity, physicochemical instability, low pharmacokinetics, low bioavailability, poor absorption, rapid metabolism, poor penetration, poor targeting efficacy, sensitivity to alkaline conditions, metal ion heat and light hinder the practical applications of curcumin which can be troubleshot by nanocurcumin formulations (Ghosh et al. 2011; Flora et al. 2013; Yallapu et al. 2012a).

Nanotization, nanoformulation (Ghosh et al. 2011) and nanoencapsulation are some nanotechnological interventions which make curcumin "a magic bullet" due to high surface area coming into contact with the solvent medium (Gera et al. 2017). Nanotization leads to enhanced solubility, dispersity, site-specific delivery, controlled release and enhanced cellular uptake (Zielinska et al. 2020). Nanoformulated curcumin shows threefold increase in bioavailability, increased accumulation at the target site, increased circulatory half-life due to decreased susceptibility for protein adsorption (opsonization) (Setthacheewakul et al. 2010).

Common techniques used to prepare nanocurcumin are nanoprecipitation, single emulsion, microemulsion, spray drying, emulsion polymerization, solvent evaporation, antisolvent precipitation, ultrasonication, coacervation technique, ionic gelation, wet milling, solid dispersion, thin-film hydration and Fessi method. The two most popular and effective techniques however are ionic gelation and antisolvent precipitation. Different forms of nanocurcumin and their advantages over the bulk curcumin are listed in Table 6.4.

The pharmacokinetics and the therapeutic applications of curcumin can be enhanced when curcumin is integrated (i.e., conjugated /encapsulated) with polymeric nanoparticles, solid-lipid nanoparticles, magnetic nanoparticles, gold nanoparticles and albumin-based nanoparticles. Natural and synthetic polymers including N-isopropylacrylamide (NIPAAM), polyvinyl alcohol (PVA), poly(lactic-co-glycolic

Table 6.4 Nanocurcumin and their value-added properties

Nanocurcumin	Advantages	References
Liposomal curcumin with targeting antibody as a ligand	Site-specific delivery	Thangapazham et al. (2008)
Encapsulation in polyethylene glycol and polycaprolactone	Effective sustained delivery and enhanced bioavailability	Sasaki et al. (2011)
PEG and PLGA nanoparticles loaded with curcumin	Improved cellular uptake and pharmacological effect	Garodia et al. (2007)
Curcumin loaded to solid-lipid nanoparticles	High stability and compatibility	Jourghanian et al. (2016)
Curcumin loaded to albumin nanoparticles with folate ligand	Sustained release at target site. Prolonged retention time Possibility of intravenous administration	Song et al. (2016)
Curcumin–cyclodextrin nanosponges	High solubility Prolonged control release Non-hemolytic	Darandale and Vavia (2013)
Curcumin–chitin nanogels	High stability Sustained release	Mangalathillam et al. (2012)
Curcumin–gold nanoconjugate	Extended blood circulation, high stability and versatile delivery of curcumin	Elbialy et al. (2019)
Curcumin-coated magnetic nanoparticles	Targeting inflammatory cells Bioimaging	Ayubi et al. (2019) Bandari et al. (2016)

acid) (PLGA), polyethylene glycol monoacrylate [NIPAAM (VP/PEG A)], N-vinyl-2-pyrrolidone, silk fibroin, hydrophobically modified starch and chitosan have been successfully utilized for synthesis of curcumin nanoparticles owing to their enhanced cellular uptake (Shome et al. 2016). Liposomal carriers solubilize the curcumin in the phospholipidic bilayer, increase the biological and pharmacological effects (Chang et al. 2018) and show polydispersity index and high encapsulation efficiency (Huang et al. 2019). Curcumin in solid-lipid nanoparticles shows extended cellular uptake with improved dispersibility, chemical stability, biocompatibility and lower toxicity (Fathy Abd-Ellatef et al. 2020). The anti-inflammatory activity of curcumin was enhanced after nanotechnological intervention (Al-Rohaimi 2015). The enhanced anti-inflammatory activity of cyclodextrin–curcumin complex as compared to free curcumin is attributed to the high affinity of the former to the inflammatory transcription factors such as nuclear factor kappa B which leads to higher degree of inhibition (Yadav et al. 2010). Iron oxide nanoparticle conjugated with curcumin alters the potential of the mitochondrial membrane (Yallapu et al. 2012b).

6.22.6.1 Nanocurcumin as Antioxidant

Curcumin nanorods prepared by precipitation method possess antioxidant and antitoxic activity in rat model at a dose lower than the native curcumin (Devasena and Rajasekar 2015). Curcumin nanorods exhibit excellent antioxidant activity by alleviating oxidative stress-induced lipid peroxidation and by upregulating enzymatic and non-enzymatic antioxidants. Curcumin nanosuspensions in Tween 80 are highly soluble in aqueous medium as compared to free curcumin retaining the structure and the antioxidant activity (Carvalho et al. 2015). Liposomal curcumin inhibits oxidative stress and enhances antioxidants in cells (Dogaru et al. 2020). Alginate–curcumin nanoparticles, curcumin nanocrystal and nanoprecipitated curcumin nanoparticles exhibit potent antioxidant activity (Siddique et al. 2013; Ranjbar et al. 2020; Yen et al. 2010).

6.22.6.2 Immunomodulatory Activity of Nanocurcumin

Nanocurcumin formulations inhibit the pro-inflammatory cytokines (TNF-α, IL-1β and MIP-1α) and increase the natural killer cells' activity and phagocytosis (macrophage) to a greater extent when compared to curcumin. Moreover, the nanoformulation is non-toxic and it can be used against various inflammatory conditions including the cytokine storm of SARS CoV2 infection. Nanocurcumin encapsulated in liposomes exhibits better anti-inflammatory activity than the free curcumin by decreasing TNF-α and matric metalloproteases (MMP-2 and MMP-9) (Trivedi et al. 2017; Dogaru et al. 2020).

6.22.6.3   Antiviral Nanocurcumin

Curcumin nanoformulation exhibited threefold higher antiviral activity by interfering with the expression of inflammatory cytokines IL-1β, topo II α and COX-
2 (Gandapu et al. 2011). Curcumin-modified silver nanoparticles inhibit respiratory syncytial virus (RSV) infection with a significant reduction of the viral loads
without side effects (Yang et al. 2016). Curcumin nanomicelles have the ability to
inhibit the binding and cellular entry of positive-sense RNA virus, the hepatitis C
virus (Naseri et al. 2017). Nanocurcumin formulation is capable of increasing the
membrane integrity (Luer et al. 2012).

Some of the nanocurcumin formulations with potential for combating the SARS
CoV2 infection have been patented: liposomal curcumin, chitosan-encapsulated
curcumin (Kurzrock et al. 2011; Kumar et al. 2009; Kumar et al. 2012), curcumin
nanoemulsions (Khamar et al. 2013; Pathak and Tran 2012), curcumin–cyclodextrin conjugate (Yallapu et al. 2010), curcumin-loaded magnetic nanoparticles and
acidic sophorolipid encapsulated curcumin (Chauhan et al. 2013). Apart from therapeutic applications, curcumin-loaded Fe3O4-magnetic nanoparticles show excellent
cellular uptake and it is used for imaging applications (Aeineh et al. 2018).

On the whole, the anti-inflammatory, antioxidant, antiviral and immunomodulatory activities discussed in this section suggest that nanocurcumin and nanocurcumin
analogs may evolve as a promising candidate for fighting SARS CoV2 infection.
As most of the studies are at the proof of concept stage, many clinical studies are
warranted for the authenticated use of nanocurcumin for COVID-19 treatment.

References

Abd Ellah NH, Tawfeek HM, John J, Hetta HF. Nanomedicine as a future therapeutic approach for
    Hepatitis C virus. Nanomedicine. 2019;14(11):1471–91.
Abraham J. Passive antibody therapy in COVID-19. Nature Rev Immunol 2020; 20(7): 401–403.
    https://doi.org/10.1038/s41577-020-0365-7.
Aeineh N, Salehi F, Akrami M, Nemati F, Alipour M, Ghorbani M, et al. Glutathione conjugated
    polyethylenimine on the surface of Fe3O4 magnetic nanoparticles as a theranostic agent for
    targeted and controlled curcumin delivery. J Biomater Sci PolymerEdn. 2018;29:1109–25.
Aggarwal BB, Yuan W, Li S, Gupta SC. Curcumin-free turmeric exhibits anti-inflammatory
    and anticancer activities: identification of novel components of turmeric. Mol Nutr Food Res.
    2013;57(9):1529–42.
Ahmadpour E, Godrati-Azar Z, Spotin A, Norouzi R, Hamishehkar H, Nami S, Heydarian P. Nanos-
    tructured lipid carriers of ivermectin as a novel drug delivery system in hydatidosis. Parasites
    Vectors. 2019;12:469.
Ahmed EM, Solyman SM, Mohamed N, Boseila AA, Hanora A. Antiviral activity of Ribavirin
    nanoparticles against measles virus. Cell Mol Biol (Noisy-Le-Grand). 2018;64(9):24–32.
Aleksandrovich KA, Aleksandrovich BI, Sergeevich NK, Vladimirovna MY. Method for producing
    nanocapsules of umifenovir (arbidol) in carrageenan. 2016; Patent No: RU0002599885. https://
    patentscope.wipo.int/search/en/detail.jsf?docId=RU179507726&tab=NATIONALBIBLIO.
Alexander W. Tarr,1,2 Pierre Lafaye,3 Luke Meredith,4 Laurence Damier-Piolle,5,6 Richard
    A. Urbanowicz,1,2 Annalisa Meola,5,6 Jean-Luc Jestin,5,6 Richard J. P. Brown,1,2 Jane A.
    McKeating,4 Felix A. Rey,5,6 Jonathan K. Ball,1,2 and Thomas Krey5. An Alpaca Nanobody
    Inhibits Hepatitis C Virus Entry and Cell-to-Cell Transmission. Hepatology 2013;58 (3):932–939.

Al-Rohaimi AH. Comparative anti-inflammatory potential of crystalline and amorphous nano curcumin in topical drug delivery. J Oleo Sci. 2015;64(1):27–40.

Amaravadi RK, Lippincott-Schwartz J, Yin XM, Weiss WA, Takebe N, Timmer W, DiPaola RS, Lotze MT, White E. Principles and current strategies for targeting autophagy for cancer treatment. Clin Cancer Res. 2011;17(4):654–66.

Antalis TM, Bugh TH, Wu Q. Membrane-anchored serine proteases in health and disease. Prog Mol Biol Transl Sci. 2011;99:1–50.

Archakov AI, Guseva MK, Uchaykin VF, Ipatova OM, Dochshitsin YF et al. Pharmaceutical composition containing arbidol in the form of phospholipid nanoparticles. 2012; Patent No: EP2431038A1. https://patents.google.com/patent/EP2431038A1/en.

Arya R, Das A, Prashar V, Kumar M. Potential inhibitors against papain-like protease of novel coronavirus (SARS-CoV-2) from FDA approved drugs. ChemrxivOrg. 2020. https://doi.org/10.26434/chemrxiv.11860011.v2.

Atala B. Jena AB, Kanungo N, Nayak V, Chainy GBN, Dandapet J. Catechin and Curcumin interact with corona (2019-nCoV/SARS-CoV2) viral S protein and ACE2 of human cell membrane: insights from computational study and implication for intervention. Pharmacodynamics. 2020; Preprint. doi: https://doi.org/10.21203/rs.3.rs-22057/v1.

Aydemir D, Ulusu N. Correspondence: Angiotensin-converting enzyme 2 coated nanoparticles containing respiratory masks, chewing gums and nasal filters may be used for protection against COVID-19 infection. Travel Med Infect Dis. 2020;101697.

Ayubi M, Karimi M, Abdpour S, Rostamizadeh K, Parsa M, Zamani M, et al. Magnetic nanoparticles decorated with PEGylated curcumin as dual targeted drug delivery: synthesis, toxicity and biocompatibility study. Mater Sci Eng C. 2019;104:109810.

Bacha U, Barrila J, Velazquez-Campoy A, Leavitt SA, Freire E. Identification of novel inhibitors of the SARS coronavirus main protease 3CLpro. Biochemistry. 2004;43(17):4906–12.

Baglivo M, Baronio M, Natalini G, Beccari T, Chiurazzi P, Fulcheri E, Petralia P, Michelini S, Fiorentini G, Miggiano GA, Morresi A, Tonini G, Bertelli M. Natural small molecules as inhibitors of coronavirus lipiddependent attachment to host cells: a possible strategy for reducing SARS-COV-2 infectivity? Acta Biomedica. 2020;91(1):161–4.

Bains BK, Greene MK, McGirr LM, Dorman J, Farrow SN, Scott CJ. Encapsulation of the p38 MAPK inhibitor GSK 678361A in nanoparticles for inflammatory-based disease states. J Interdisc Nanomed. 2016;1(3):85–92.

Bajpai AK, Choubey J. Design of gelatin nanoparticles as swelling controlled delivery system for chloroquine phosphate. J Mater Sci - Mater Med. 2006;17:345–58.

Balaña-Fouce R, Reguera RM, Cubriia JC, Ordonez D. The pharmacology of leishmaniasis. Gen Pharmacol. 1998;30:435–43.

Bellido E, Hidalgo T, Lozano MV, Guillevic M, Simon Vazquez R, Santander Ortigo MJ, et al. Heparin-engineered mesoporous iron metal-organic framework nanoparticles: toward stealth drug nanocarriers. Adv Healthc Mater. 2015;4(8):1246–57.

Bhandari R, Gupta P, Dziubla T, Hilt JZ. Single step synthesis, characterization and applications of curcumin functionalized iron oxide magnetic nanoparticles. Mater Sci Eng C. 2016;67:59–64.

Billinger E, Zuo S, Johansson G. characterization of serine protease inhibitor from solanum tuberosum conjugated to soluble dextran and particle carriers. ACS Omega. 2019;4(19):18456–64.

Bittmann S, Moschuring Alieva E, Weissenstein A, Luchter E, Villalon G. The role of TMPRSS2-inhibitor camostat in the pathogenesis of COVID-19 in lung cells. Biomed J Sci Tech Sci Res. 2020;20(3):20875–6.

Bosch-Barrera J, Martin-Castillo B, Buxo Pujolras M, Joan B, José E, Javier M. Silibinin and SARS-CoV-2: dual targeting of host cytokine storm and virus replication machinery for clinical management of COVID-19 patients. J Clin Med. 2020;9:1770.

Caly L, Druc J, Catton MG, Jans DA, Wagstaff KM. The FDA-approved Drug Ivermectin inhibits the replication of SARS-CoV-2 in vitro. Antiviral Res. 2020;178:104787.

Carvalho DDM, Takeuchi KP, Geraldin RM, De Mouza CJ. Production, solubility and antioxidant activity of curcumin nanosuspension. Food Sci Technol. 2015;35(1):115–9.

Cavezzi A, Troiani E, Corrao S. COVID-19: hemoglobin, iron, and hypoxia beyond inflammation. A narrative review. . Clinics and Practice. 2020;10(2):1271.

Chang M, Wu M, Li H. Antitumor activities of novel glycyrrhetinic acid-modified curcumin-loaded cationic liposomes in vitro and in H22 tumor-bearing mice. Drug Delivery. 2018;25:1984–95.

Chauhan S, Jaggi M, Yallapu MM. Magnetic nanoparticle formulations, methods for making such formulations, and methods for their use. 2013; US Patent Number: US 20130245357A1.

Chenthamara D, Subramaniam S, Ramakrishnan SG, Krishnaswamy S, Essa MM, Lin FH, Oronfleh MW. Therapeutic efficacy of nanoparticles and routes of administration. Biomate Res. 2019;23:20.

Chen TY, Chen DY, Wen HW, Ou JL, Chiou SS, Cheng JM, Wong ML, Hsu WL. Inhibition of enveloped viruses infectivity by curcumin. PLoS ONE. 2013;8(5):e62482.

Chen X, Chou CY, Chang GG. Thiopurine analogue inhibitors of severe acute respiratory syndrome-coronavirus papain-like protease, a deubiquitinating and deISGylating enzyme. Antiviral Chem Chemother. 2009;19:151–6.

Chen R, Cheng G, So MH, WU J, Lu Z, Che CM, Sun H. Bismuth subcarbonate nanoparticles fabricated by water-in-oil microemulsion-assisted hydrothermal process exhibit anti- Helicobacter pylori properties. Mater Res Bull. 2010;45(5):654–658.

Chen W, Feng P, Liu K, Wu M, Lin H. Computational identification of small interfering RNA targets in SARS-CoV-2. Virol Sin. 2020;35(3):359–61.

Cheng Y, Wong R, Soo YO, Wong WS, Lee CK, Ng MH, Chan P, Wong KC, Leung CB, Cheng G. Use of convalescent plasma therapy in SARS patients in Hong Kong. Eur J Clin Microbiol Infect Dis 2005;24:44–46.

Chiodo F, Bruijns SCM, Rodriguez E, Li E, Molinaro A, Silipo A, Lorenzo FD, Gracia Riveria D, Babin YV, Verez-Bencomo V, van Kooyk Y. Novel ACE2-independent carbohydrate-binding of SARS-CoV-2 spike protein to host lectins and lung microbiota. 2020;1–11. bioRxiv preprint. doi: https://doi.org/https://doi.org/10.1101/2020.05.13.092478.

Choi KS, Aizaki H, Lai MMC. Murine coronavirus requires lipid rafts for virus entry and cell–cell fusion but not for virus release. J Virol. 2005;79:9862–71.

Cho JK, Curtis-Long MJ, Lee KH, Kim DW, Ryu HW, Yuk HJ, Park KH. Geranylated flavonoids displaying SARS-CoV papain-like protease inhibition from the fruits of Paulownia tomentosa. Bioorg Med Chem. 2013;21:3051–7.

Cilurzo F, Di Marzio L, Carafa M, Ventura CA, Wolfram J, Paolino D, Celia C. Liposomal chemotherapeutics. Future Oncol. 2013;9:1849–59.

Colson P, Rolain JM, Lagier JC, Brouqui P, Raoult D. Chloroquine and hydroxychloroquine as available weapons to fight COVID-19. Int J Antimicrob Agents. 2020;55(4):105932.

Craparo EF, Drago SE, Mauro N, Giammona G, Cavallaro G. Design of new polyaspartamide copolymers for siRNA delivery in antiasthmatic therapy. Pharmaceutics. 2020;12(2):89.

Darandale SS, Vavia PR. Cyclodextrin-based nanosponges of curcumin: formulation and physico-chemical characterization. J Incl Phenom Macrocycl Chem. 2013;75:315–22.

Datta A, Brosh RM Jr. New insights into DNA helicases as druggable targets for cancer therapy. Front Mol Biosci. 2018;5. doi: https://doi.org/10.3389/fmolb.2018.00059.

David L, Maria F, Ana C, Josefa B, Laura B, Lluisa BC. Multifunctional serine protease inhibitor-coated water-soluble gold nanoparticles as a novel targeted approach for the treatment of inflammatory skin diseases. Bioconjug Chem. 2018;29(4):1060–72.

Decroly E, Imbert I, Coutard B, Bouvet M, Selisko B, Alvarez K, et al. Coronavirus non-structural protein 16 is a cap-0 binding enzyme possessing (nucleoside-2'O)-methyltransferase activity. J Virol. 2008;82(16):8071–84.

Devasena T, Rajasekar A. Of bulk and nano: comparing the hepatoprotective efficacy of curcumin in rats. Chem Sci Rev Lett. 2014;3(12):951–6.

Devasena T, Rajasekar A. Facile synthesis of curcumin nanocrystals and validation of its antioxidant activity against circulatory toxicity in wistar rats. J Nanosci Nanotech. 2015;15(6):4119–4125

Devasena T, Rajasekaran KN, Menon VP. Bis-1,7-(2-hydroxyphenyl)-hepta-1,6-diene-3,5-dione (a curcumin analog) ameliorates DMH-induced hepatic oxidative stress during colon carcinogenesis. Pharmacol Res. 2002;46(1):39–45.

Devaux CA, Rolain JM, Colson P, Raoult D. New insights on the antiviral effects of chloroquine against coronavirus: what to expect for COVID-19? Int J Antimicrob Agents. 2020;50(5):105938.

Dey P, Bergmann T, Cuellar-Camacho JL, Ehrmann S, Chowdhury MS, Zhang M, Dahmani I, Haag R, Azab W. multivalent flexible nanogels exhibit broad-spectrum antiviral activity by blocking virus entry. ACS Nano. 2018;12(7):6429–42.

Ding L, Zhang XX, Wei P, Fan K, Lai L. The interaction between severe acute respiratory syndrome coronavirus 3C-like proteinase and a dimeric inhibitor by capillary electrophoresis. Anal Biochem. 2005;343:159–65.

Dogaru G, Bulboaca AE, Gheban D, Boarescu PM, Rus V, Festila D, et al. Effect of liposomal curcumin on acetaminophen hepatotoxicity by down-regulation of oxidative stress and matrix metalloproteinases. Vivo. 2020;34(2):569–82.

Dormont F, Brusini R, Cailleau C, Reynaud F, Peramo A, Gendron A, Mougin J, Gaudin F, Varna M, Couvreur P. Squalene-based multidrug nanoparticles for improved mitigation of uncontrolled inflammation. Sci Adv. 2020;6(23):eaaz5466.

Dou DD. Preparation of ivermectin solid lipid nanoparticles and preliminary study on transdermal properties. Artif Cell Nanomed Biotechnol. 2016;46:255–62.

Duan K, Liu B, Li C, et al. Effectiveness of convalescent plasma therapy in severe COVID-19 patients. Proc Nat Acad Sci USA 2020;117:9490–9496.

Duarah S, Durai RD, Narayanan VB. Nanoparticle-in-gel system for delivery of vitamin C for topical application. Drug Deliv Transl Res. 2017;7(5):750–60.

Elbialy NS, Abdelfatah EA, Khalil WA. Antitumor activity of curcumin-green synthesized gold nanoparticles: in vitro study. BioNanoScience. 2019;9:813–20. https://doi.org/10.1007/s12668-019-00660-w.

Fatemian T, Moghimi HR, Chowdhury EH. Intracellular delivery of siRNAs targeting AKT and ERBB2 genes enhances chemosensitization of breast cancer cells in a culture and animal model. Pharmaceutics. 2019;11:458

Fathy Abd-Ellatef GE, Gazzano E, Chirio D, Hamed AR, Belisario DC, Zuddas C, et al. Curcumin-loaded solid lipid nanoparticles bypass P-glycoprotein mediated doxorubicin resistance in triple negative breast cancer cells. Pharmaceutics. 2020;12(2):96.

Fayed BE, Tawfik AF, Yassin AE. Novel erythropoietin-loaded nanoparticles with prolonged in vivo response. J Microencapsul. 2012;29(7):650–6.

Flora G, Gupta D, Tiwari A. Nanocurcumin: a promising therapeutic advancement over native curcumin. Crit Rev Ther Drug Carrier Syst. 2013;30(4):331–68.

Francis AP, Ganapathy S, Palla VR, Murthy PB, Ramaprabhu S, Devasena T. One time 207 nose-only inhalation of MWCNTs: exploring the mechanism of toxicity by intermittent sacrifice in Wistar rats. Toxicol Rep. 2015;2:111–20.

Francis AP, Devasena T, Selvam G, Rajsekhar Palla V, BalakrishnaMurthy P, Ramaprabhu S. Multi-walled carbon nanotube-induced inhalation toxicity: Recognizing nano bis-demethoxy curcumin analog as an ameliorating candidate. Nanomed Nanotechnol Biol Med. 2018;14(6):1809–1822.

Frieman M, Yount B, Heise M, Kopecky-Bromberg SA, Palese P, Baric RS. Severe acute respiratory syndrome coronavirus ORF6 antagonizes STAT1 function by sequestering nuclear import factors on the rough endoplasmic reticulum/Golgi membrane. J Virol. 2007;81(18):9812–24.

Gagne JF, Désormeaux A, Perron S, Tremblay MJ, Bergeron MG. Targeted delivery of indinavir to HIV-1 primary reservoirs with immunoliposomes. Biochim Biophys Acta 2002;1558:198–210.

Gandapu U, Chaitanya R, Kishore G, Reddy RC, Kondapi AK. Curcumin-loaded apotransferrin nanoparticles provide efficient cellular uptake and effectively inhibit HIV-1 replication in vitro. PLoS ONE. 2011;6:e23388.

Gao J, Tian Z, Yang X. Breakthrough: chloroquine phosphate has shown apparent efficacy in treatment of COVID-19 associated pneumonia in clinical studies. Biosci Trends. 2020;14(1):72–3.

Garodia P, Ichikawa H, Malani N, Sethi G, Aggarwal BB. From ancient medicine to modern medicine: ayurvedic concepts of health and their role in inflammation and cancer. J Soc Integr Oncol. 2007;5(1):25–37.

Garrido C, Simpson CA, Dahl NP, Bresee J, Whitehead DC, Lindsey EA, Harris TL, Smith CA, Carter CJ, Feldheim DL, Melander C, Margolis DM. Gold nanoparticles to improve HIV drug delivery. Future Med Chem. 2015;7(9):1097–107.

Gassen NC, Niemeyer D, Muth D. Corman VM, Martinelli S, Gassen A et al. SKP2 attenuates autophagy through Beclin1-ubiquitination and its inhibition reduces MERS-coronavirus infection. Nat Commun. 2019;10:5770.

Gera M, Sharma N, Ghosh M, Huynh DL, Lee SJ, Min T, Kwon T, Jeong DK. Nanoformulations of curcumin: an emerging paradigm for improved remedial application. Oncotarget. 2017;8(39):66680–98.

Ghaffari H, Tavakoli A, Moradi A, et al. Inhibition of H1N1 influenza virus infection by zinc oxide nanoparticles: another emerging application of nanomedicine. J Biomed Sci. 2019;26(1):70.

Ghosh M, Singh AT, Xu W, Sulchek T, Gordon LI, Ryan RO. Curcumin nanodisks: formulation and characterization. Nanomed Nanotechnol Biol Med. 2011;7:162–167.

Ghosh S, Firdous SM, Nath A. siRNA could be a potential therapy for COVID-19. EXCLI J. 2020;19:528–31.

Grabowska-Jadach I, Drozd M, Kulpińska D, Komendacka K, Pietrzak M. Modification of fluorescent nanocrystals with 6-thioguanine: monitoring of drug delivery. Appl Nanosci. 2020;10:83–93.

Greene MK, Richards DA, Nogueira JCF, Campbell K, Smyth P, Fernandez M, Scott CJ, Chudasama V. Forming next-generation antibody–nanoparticle conjugates through the oriented installation of non-engineered antibody fragments. Chem Sci. 2018;9:79–87.

Grimes JM, Grimes KV. P38 MAPK inhibition: a promising therapeutic approach for COVID-19. J Mol Cell Cardiol. 2020;144:63–5.

Guo D, Dou D, Li X, Zhang Q, Bhutto ZA, Wang L. Ivermectin-loaded solid lipid nanoparticles: preparation, characterisation, stability and transdermal behaviour. Artif Cells Nanomed Biotechnol. 2018;46(2):255–62.

Gyebi GA, Ogunro OB, Adegunloye AP, Ogunyemi OM, Afolabi SO. Potential inhibitors of coronavirus 3-chymotrypsin-like protease (3CLpro): an in silico screening of alkaloids and terpenoids from African medicinal plants. J Biomol Struct Dyn. 2020;1–13. https://doi.org/https://doi.org/10.1080/07391102.2020.1764868.

Halder A, Das S, Ojha D, Chattopadhyay D, Mukherjee A. Highly monodispersed gold nanoparticles synthesis and inhibition of herpes simplex virus infections. Mater Sci Eng C. 2018;89:413–21.

Hamers-Casterman C, Atarhouch T, Muyldermans S, Robinson G, Hamers C, Songa EB, Bendahman N, Hamers R. Naturally occurring antibodies devoid of light chains. Nature 1993;363:446–448.

Han GG, Chang CG, Juo HJ, Lee SH, Yeh JT, Hsu X. ChenPapain-like protease 2 (PLP2) from severe acute respiratory syndrome coronavirus (SARS-CoV): expression, purification, characterization, and inhibition. Biochemistry. 2005;44:10349–59.

Hattori Y, Nakamura M, Takeuchi N, Tamaki K, Ozaki K, Onishi H. Effect of cationic lipid type in PEGylated liposomes on siRNA delivery following the intravenous injection of siRNA lipoplexes. World Acad Sci J. 2019;1:74–85.

Heidary F, Gharebaghi R. Ivermectin: a systematic review from antiviral effects to COVID-19 complementary regimen. J Antibiot. 2020;73:593–602.

Hoffmann M, Weber KH, Schroeder S, Kruger N, Herrler T, Erichsen S, Schieger TS, et al. SARS-CoV-2 cell entry depends on ACE2 and TMPRSS2 and is blocked by clinically proven protease inhibitor. Cell. 2020;181:1–10.

Huang J, Song W, Huang H, Sun Q. Pharmacological therapeutics targeting RNA-dependent RNA polymerase, proteinase and spike protein: from mechanistic studies to clinical trials for COVID-19. J Clin Med. 2020;9(4):1131.

Huang M, Liang C, Tan C, Huang S, Ying., Wang Y, Wang Z et al. Liposome co-encapsulation as a strategy for the delivery of curcumin and resveratrol. Food Funct. 2019;10:6447–6458.

Hu CJ, Chang WS, Fang ZS, Cheng YT, Wang WL, Tsai HH, et al. Nanoparticulate vacuolar ATPase blocker exhibits potent host-targeted antiviral activity against feline coronavirus. Sci Rep. 2017;7(1):13043.

Hu TY, Frieman M, Wolfram J. Insights from nanomedicine into chloroquine efficacy against COVID-19. Nat Nanotechnol. 2020;15:247–9.

Iacobellis G. COVID-19 and diabetes: can DPP4 inhibition play a role? Diabetes Res Clin Pract. 2020;162:108125.

Jacobs J, Grum-Tokars V, Zhou Y, Turlington M, Saldanha SA, Chase P, Eggler A, Dawson ES, Baez-Santos YM, Tomar S, et al. Discovery, synthesis, and structure-based optimization of a series of N-(tert-butyl)-2-(N-arylamido)-2-(pyridin-3-yl) acetamides (ML188) as potent noncovalent small molecule inhibitors of the severe acute respiratory syndrome coronavirus (SARS-CoV) 3CL protease. J Med Chem. 2013;56:534–46.

Jahangirian H, Kalantari K, Izadiyan Z, Rafiee-Moghaddam R, Shameli K, Webster TJ. A review of small molecules and drug delivery applications using gold and iron nanoparticles. Int J Nanomed. 2019;14:1633–57.

Jang Y, Shin JS, Yoon YS, Go YY, Lee HW, Kwon OS, Park S, Park MS, Kim M. Salinomycin inhibits influenza virus infection by disrupting endosomal acidification and viral matrix protein 2 function. J Virol. 2018;92(24):e0144118.

Jaruga E, Salvioli S, Dobrucki J, Chrul S, Bandorowicz-Pikula J, Sikora E, et al. Apoptosis-like, reversible changes in plasma membrane asymmetry and permeability, and transient modifications in mitochondrial membrane potential induced by curcumin in rat thymocytes. FEBS Lett. 1998;433:287–93.

Jin L, Wang Q, Chen J, Wang Z, Xin H, Zhang D. Efficient delivery of therapeutic siRNA by Fe3O4 magnetic nanoparticles into oral cancer cells. Pharmaceutics. 2019;11(11):17.

Joshi RS, Jagdale SS. Bansode SB, Shankar SS, Tellis MB, Pandya VK, Chugh A, Giri AP, Kulkarni MJ. Discovery of potential multi-target-directed ligands by targeting host-specific SARS-CoV-2 structurally conserved main protease. J Biomol Struct Dyn. 2020;1–16. https://doi.org/https://doi.org/10.1080/07391102.2020.1760137.

Jourghanian P, Ghaffari S, Ardjmand M, Haghighat S, Mohammadnejad M. Sustained release curcumin loaded solid lipid nanoparticles. Adv Pharm Bull. 2016;6(1):17–21.

Kandeel M, Al-Taher A, Park BK, Kwon HJ, Al-Nazawi M. A pilot study of the antiviral activity of anionic and cationic polyamidoamine dendrimers against the Middle East respiratory syndrome coronavirus. J Med Virol. 2020;1–6. doi: https://doi.org/10.1002/jmv.25928.

Kang B, Mackey MA, El-Sayed MA. Nuclear targeting of gold nanoparticles in cancer cells induces DNA damage, causing cytokinesis arrest and apoptosis. J Am Chem Soc. 2010;132:1517–9.

Karbalaei M, Keikha M. Curcumin as an Herbal Inhibitor Candidate Against HTLV-1 Protease. Jentashapir J. Health Res. 2019;10(1).

Kaur G, Narang RK, Rath G, Goyal AK. Advances in pulmonary delivery of nanoparticles. Artif Cells Blood Substit Biotechnol. 2012;40(1–2):75–96.

Keyaertse, Vijgen L, Maes P: In vitro inhibition of severe acute respiratory syndrome coronavirus by chloroquine. Biochemical and Biophysical Research Communication. 2004;323(1):264–268.

Keyaerts E, Li S, Vijgen L, Rysman E, Verbeeck J, Van Ranst M, et al. Antiviral activity of chloroquine against human coronavirus OC43 infection in 8 newborn mice. Antimicrob Agents Chemother. 2009;53:3416–21.

Keyaerts, E., Vijgen, L., Pannecouque, C., Van Damme, E., Peumans, W., Egberink, H., Balzarini, J., & Van Ranst, M. Plant lectins are potent inhibitors of coronaviruses by interfering with two targets in the viral replication cycle. Antiviral Res 2007;75(3):179–187. https://doi.org/10.1016/j.antiviral.2007.03.003.

Khamar BM, Gogia AP, Goda CC, Shenoy DB, Shrivastava RR, Patravale VB et al. Pharmaceutical compositions of curcumin. 2013;U.S Patent. Number: 9474727B2.

Khan RJ, Jha RK, Amera GM, Jain M, Singh E, Pathak A, Singh RP, Muthukumaran J, Singh AK. Targeting SARS-CoV-2: a systematic drug repurposing approach to identify promising inhibitors against 3C-like proteinase and 2'-O-ribose methyltransferase. J Biomol Struct Dyn. 2020;2:1–14.

Kim JM, Chung YS, Jo HJ, Lee NJ, Kim MS, Woo SH, Park S, Kim JW, Kim HM, Han MG. Identification of Coronavirus Isolated from a Patient in Korea with COVID-19. Osong Publ Health Res Perspect. 2020;11:3–7.

Kitamura K, Tomita K. Proteolytic activation of the epithelial sodium channel and therapeutic application of a serine protease inhibitor for the treatment of salt-sensitive hypertension. Clin Exp Nephrol. 2012;16(1):44–8.

Klimkowska BM, Poplawska M, Grudzinski IP. Nanocomposites as biomolecules delivery agents in nanomedicine. J Nanobiotechnol. 2019;17:48. https://doi.org/10.1186/s12951-019-0479-x.

Ko JH, Seok H, Cho SY, Ha YE, Baek JY, Kim SH, Kim YJ, Park JK, Chung CR, Kang ES, Cho D, Müller MA, Drosten C, Kang CI, Chung DR, Song JH, Peck KR. Challenges of convalescent plasma infusion therapy in Middle East respiratory coronavirus infection: a single centre experience. Antiviral therapy 2018;23(7):617–622.

Knudsen NO, Ronholt S, Salte RD, Jorgensen L, Thormann T, Basse LH, Hansen J, Frokjaer S, Foged C. Calcipotriol delivery into the skin with PEGylated liposomes. Euro J Pharm Biopharm. 2012;81(3):532–9.

Kumar A, Mohapatra SS, Cameron DF. Nanoparticle targeted drug delivery to the lungs using extra-testicular sertoli cells. 2009; WO Patent No: 2009105278A2.

Kumar KS, Gnanaprakash D, Mayilvaganan K, Arunraj C, Mohankumar S. Chitosan-gold nanoparticles as delivery systems for curcumin. Int J Pharm Sci Res. 2012;3:4533.

Kurzrock R, Li L, Mehta K, Aggarawal BB. Liposomal curcumin for treatment of cancer. 2011; United States Patent No: US20060067998A1.

Kuzmov A, Minko T. Nanotechnology approaches for inhalation treatment of lung diseases. J Control Release. 2015;219:500–18.

Lalloz A, Bolzinger MA, Faivre J, Latreille PL, Garcia Ac A, Rakotovao C, Rabanel JM, Hildgen P, Banquy X, Briançon S. Effect of surface chemistry of polymeric nanoparticles on cutaneous penetration of cholecalciferol. Int J Pharm. 2018;553:120–31.

Lang J, Yang N, Deng J, Liu K, Yang P, Zhang G, Jiang C. Inhibition of SARS pseudovirus cell entry by lactoferrin binding to heparan sulfate proteoglycans. PLoS ONE. 2011;6(8):e23710.

Lee C, Lee JM, Lee NR, Kim DE, Chong Y. Investigation of the pharmacophore space of severe acute respiratory syndrome coronavirus (SARS-CoV) NTPase/helicase by dihydroxychromone derivatives. Bioorg Med Chem Lett. 2009;19:4538–41.

Lee H, Lee MY, Bhang SH, Kim BS, Kim YS, Ju JH, Kim KS, Hahn SK. Hyaluronate–gold nanoparticle/tocilizumab complex for the treatment of rheumatoid arthritis. Acs Nano. 2014 May 27;8(5):4790–8.Lee H, Lee MY, Bhang SH, Kim BS, Kim YS, Ju JH, Kim KS, Hahn SK. Hyaluronate–gold nanoparticle/tocilizumab complex for the treatment of rheumatoid arthritis. Acs Nano. 2014;8(5):4790–8.

Lembo D, Trotta F, Cavalli R. Cyclodextrin-based nanosponges as vehicles for antiviral drugs: challenges and perspectives. Nanomedicine. 2018;13(5):477–80.

Lima T, Feitosa RC, Dos Santos-Silva E, Dos Santos-Silva AM, Siqueira, E, Machado P, Cornélio AM, do Egito E, Fernandes-Pedrosa MF, Farias K, da Silva-Júnior AA. improving encapsulation of hydrophilic chloroquine diphosphate into biodegradable nanoparticles: a promising approach against herpes virus simplex-1 infection. Pharmasceutics. 2018;10(4):255.

Lindner HA, Fotouhi- Ardakani N, Lytvyn V, Lachance P, Sulea T, Menard R. The Papain-like protease from the severe acute respiratory syndrome coronavirus is a deubiquitinating enzyme. J Virol. 2005;79:15199–208.

Liu Z, Ying Y. The Inhibitory effect of curcumin on virus-induced cytokine storm and its potential use in the associated severe pneumonia. Front Cell Dev Biol. 2020;8:479.

Liu C, Wan L, Wang H, Zhang S, Ping Y, Cheng Y. A boronic acid–rich dendrimer with robust and unprecedented efficiency for cytosolic protein delivery and CRISPR-Cas9 gene editing. Sci Adv. 2019;5(6):eaaw8922.

Liu W, Zhang S, Nekhai., Liu S. depriving iron supply to the virus represents a promising adjuvant therapeutic against viral survival. Curr Clin Microbiol Rep. 2020;1–7. doi: https://doi.org/10.1007/s40588-020-00140-w.

Liu L, Chopra P, Li X, Wolfert MA, Tompkins SM, Boons GJ. SARS-CoV-2 spike protein binds heparan sulfate in a length- and sequence-dependent manner. bioRxiv 2020; doi: https://doi.org/10.1101/2020.05.10.087288.

Li S, Zhao Y. preparation of melatonin-loaded zein nanoparticles using supercritical CO2 antisolvent and in vitro release evaluation. Int J Food Eng. 2017;13(11):1556–3758.

Li B, Tang Q, Cheng D, Quin C, Xie FY, Wei Q, Xu J, et al. Using siRNA in prophylactic and therapeutic regimens against SARS coronavirus in Rhesus macaque. Nat Med. 2005;11:944–51.

Li W, Hulswit RJG, Widjaja I, Raj VS, McBride R, Peng W, et al. Identification of sialic acid-binding function for the Middle East respiratory syndrome coronavirus spike glycoprotein. Proc Natl Acad Sci. 2017;114(40):E8508–17.

Loretz BF, Föger M, Werle A, Bernkop-Schnürch A. Oral gene delivery: strategies to improve stability of pDNA towards intestinal digestion. J Drug Target. 2006;14(5):311–9.

Luer S, Troller R, Aebi C. Antibacterial and Anti-inflammatory kinetics of curcumin as a potential antimucositis agent in cancer patients. Nutr Cancer. 2012;64(7):975–81.

Luo M, Boudiera A, Clarota I, Maincenta P, Schneiderb R, Leroy P. Gold nanoparticles grafted by reduced glutathione with thiol function preservation. Colloid Interf Sci Commun. 2016;14:8–12.

Lu G, Hu Y, Wang Q, Qui J, Gao F, Li Y, Zhang Y, et al. Molecular basis of binding between novel human coronavirus MERS-CoV and its receptor CD26. Nature. 2013;500:227–31.

Lu M, Xiang D, Sun W, Yu T, Hu Z, Ding J, et al. Sustained release ivermectin-loaded solid lipid dispersion for subcutaneous delivery: in vitro and in vivo evaluation. Drug Deliv. 2017;24:622–31.

Mandal S, Prathipati PK, Belshan M, Destache C. A potential long-acting bictegravir loaded nano-drug delivery system for HIV-1 infection: a proof-of-concept study. Antiviral Res. 2019;167:83–8.

Manes S, del Real G, Martínez AC. Pathogens: raft hijackers. Nat Rev Immunol. 2003;3(7):557–68.

Mangalathillam S, Rejinold NS, Nair A, Lakshmanan VK, Nair SV, Jayakumar R. Curcumin loaded chitin nanogels for skin cancer treatment via the transdermal route. Nanoscale. 2012;4:239–50.

Maudens P, Seemayer CA, Pfefferle F, Jordan O, Allémann E. Nanocrystals of a potent p38 MAPK inhibitor embedded in microparticles: therapeutic effects in inflammatory and mechanistic murine models of osteoarthritis. J Control Release. 2018;276:102–12.

McKee DL, Sternberg A, Stange U, Laufer S, Naujokat C. Candidate drugs against SARS-CoV-2 and COVID-19. Pharmacol Res. 2020;157:104859.

Mehta P, McAuley DF, Brown M, Sanchez E, Tattersall RS, Manson JJ. COVID-19: Consider cytokine storm syndromes and immunosuppression. The Lancet. 2020;395:1033–4.

Metcalfe SM. COVID-19 lockdown: de-risking exit by protecting the lung with leukaemia inhibitory factor (LIF). Medicine in Drug Discovery. 2020;6:100043.

Metcalfe SM, Strom TB, Williams A, Fahmy TM. Multiple sclerosis and the LIF/IL-6 axis: use of nanotechnology to harness the tolerogenic and reparative properties of LIF. Nanobiomedicine. 2015;2:5.

Milad A, Zahra A, Reza M, Tahereh F, Saeed S. Nano-soldiers ameliorate silibinin delivery: a review study. Curr Drug Deliv. 2020;17(1):15–22.

Milewska A, Zarebski M, Nowak P, Stoze K, Potempa J, Pyrc K. Human coronavirus NL63 utilizes heparan sulfate proteoglycans for attachment to target cells. J Virol. 2014;88(22):13221–30.

Miller SE, Mathiasen S, Bright NA, Pierre F, Kelley BT, Kladt N, et al. CALM regulates clathrin-coated vesicle size and maturation by directly sensing and driving membrane curvature. Dev Cell. 2015;33:163–75.

Mizutani T. Signal transduction in SARS-CoV-infected cells. Ann N Y Acad Sci. 2007;1102(1):86–95.

Mizutani T, Fukushi S, Murakami M, Hirano T, Saijo M, Kurane I, Morikawa S. Tyrosine dephosphorylation of STAT3 in SARS coronavirus-infected Vero E6 cells. FEBS Lett. 2004;577:187–92.

Mizutani T, Fukushi S, Saijo M, Kurane I, Morikawa S. Phosphorylation of p38 MAPK and its downstream targets in SARS coronavirus-infected cells. Biochem Biophys Res Commun. 2004;319:1228–34.

Mokhtari PN, Ghorbani M, Mahmoodzadehc F. Smart co-delivery of 6-mercaptopurine and methotrexate using disulphide-based PEGylated-nanogels for effective treatment of breast cancer. New J Chem. 2019;43:12159–67.

Mounce BC, Cesaro T, Carrau L, Vallet T, Vignuzzi M. Curcumin inhibits Zika and chikungunya virus infection by inhibiting cell binding. Antiviral Res. 2017;142:148–57.

Moustafa ME, Amin AS, Magdi Y. Cytotoxicity of 6-mercaptopurine via loading on PVA-coated magnetite nanoparticles delivery system: a new era of leukemia therapy. J Nanomed Nanotech. 2018;9(6): ID: 1000521. 1–8.

Muyldermans S. Nanobodies: natural single-domain antibodies. Annu Rev Biochem. 2013;82:775–797.

Naqvi S, Mohiyuddin S, Gopinath P. Niclosamide loaded biodegradable chitosan nanocargoes: an in vitro study for potential application in cancer therapy. Roy Soc Open Sci. 2017;4(11):170611.

Naseri S, Darroudi M, Aryan E, Gholoobi A, Rahimi HR, Ketabi K, et al. The antiviral effects of curcumin nanomicelles on the attachment and entry of hepatitis C virus. Iran J Virol. 2017;11:29–35.

O'Driscoll CM, Bernkop-Schnürch A, Friedl J, Préat V, Jannin V. Oral delivery of non-viral nucleic acid-based therapeutics—do we have the guts for this? Euro J Pharm Sci. 2019;133:190–204.

Ogunwuyi O, Kumari N, Smith KA, Bolshakov O, Adesina S, Gugssa A, Akala EO. Antiretroviral drugs-loaded nanoparticles fabricated by dispersion polymerization with potential for HIV/AIDS treatment. Infect Dis Res Treat. 2016;9:21–32.

Ortega JT, Serrano ML, Pujol FH, Rangel HR. Unrevealing sequence and structural features of novel coronavirus using in silico approaches: the main protease as molecular target. EXCLI J. 2020;19:400–9.

Othman N, Masarudin MJ, Kuen CY, Dasuan NA, Abdullah LC, Md JS. Synthesis and optimization of chitosan nanoparticles loaded with L-ascorbic acid and thymoquinone. Nanomaterials. 2018;8(11):920.

Ozturk B, Argin S, Ozilgen M, McClements DJ. Nanoemulsion delivery systems for oil-soluble vitamins: influence of carrier oil type on lipid digestion and vitamin D3 bioaccessibility. Food Chem. 2015;187:499–506.

Pachetti M, Marini B, Benedetti F, Giudici F, Mauro E, Storici P, Masciovecchio C, Angeletti S, Ciccozzi M, Gallo RC, et al. Emerging SARS-CoV-2 mutation hot spots include a novel RNA-dependent-RNA polymerase variant. J Transl Med. 2020;18:179.

Park JY, Jeong HJ, Kim JH, Kim YM, Park SJ, Kim D, Park KH, Lee WS, Ryu YB. Diarylheptanoids from Alnus japonica inhibit papain-like protease of severe acute respiratory syndrome coronavirus. Biol Pharm Bull. 2012;35:2036–42.

Park YJ, Kim JH, Kim YM, Jeong HJ, Kim DW, Park KH, Kwon HJ, Park SJ, Lee WS. Ryu YB. Tanshinones as selective and slow-binding inhibitors for SARS-CoV cysteine proteases. Bioorg Med Chem. 2012;20:5928–35.

Pathak Y, Tran HT. Nanoemulsions containing antioxidants and other health-promoting compounds. 2012; United States Patent Number: US20120052126A1.

Pillaiyar T, Manickam M, Namasivayam V, Hayashi Y, Jung SH. An overview of severe acute respiratory syndrome-coronavirus (SARS-CoV) 3CL protease inhibitors: peptidomimetics and small molecule chemotherapy. J Med Chem. 2016;59(14):6595–628.

Pindiprolu SS, Phani Kumar CS, Kumar Golla VS, Lithika P, Shreyas Chanda K, Esub Basha SK, Ramachandra RK. Pulmonary delivery of nanostructured lipid carriers for effective repurposing of salinomycin as an antiviral agent. Med Hypotheses. 2020;143:109858.

Pindiprolu SKS, Chintamaneni K, Krishnamurthy PT, Sree Ganapathineedi KR. Formulation-optimization of solid lipid nanocarrier system of STAT3 inhibitor to improve its activity in triple negative breast cancer cells. Drug Dev Ind Pharm. 2019;45:304–13.

Prachi J, Soumyananda C, Ramirez-Vick, Jaime E, Ansari ZA, Virendra S, , Pinak C, Surinder SP. The anticancer activity of chloroquine-gold nanoparticles against MCF-7 breast cancer cells. Colloids Surf B. 2012;95:195–200.

Prasad S, Tyagi AK. Curcumin and its analogues: a potential natural compound against HIV infection and AIDS. Food Funct. 2015;6:3412–9.

Priya G, Saranya IH, Kiran T, Marc G. Mitochondria targeted viral replication and survival strategies—prospective on SARS-CoV-2. Frontiers Pharm. 2020;11:1364.

Quinton LJ, Mizgerd JP, Hilliard KL, Jones MR, Kwon CY, Allen E. Leukemia inhibitory factor signaling is required for lung protection during pneumonia. J Immunol. 2012;188(12):6300–8.

Ramezanli T, Kilfoyle BE, Zhang Z, Michniak-Kohn BB. Polymeric nanospheres for topical delivery of vitamin D3. Int J Pharm. 2017;516(1–2):196–203.

Ranjbar A, Gholami L, Ghasemi H, Kheiripour N. Effects of nano-curcumin and curcumin on the oxidant and antioxidant system of the liver mitochondria in aluminum phosphide-induced experimental toxicity. Nanomed J. 2020;7:58–64.

Ruan Z, Liu C, Guo Y, He Z, Huang X, Jia X, Yang T. Potential inhibitors targeting RNA-dependent RNA polymerase activity (NSP12) of SARS-CoV-2. Preprints 2020, 2020030024. doi: https://doi.org/10.20944/preprints202003.0024.v1.

Salmaso S, Caliceti P. Stealth properties to improve therapeutic efficacy of drug nanocarriers. J Drug Deliv. 2013;2013:1–19.

Sarangi B, Jana U, Sahoo J, Mohanta GP, Manna PK. Systematic approach for the formulation and optimization of atorvastatin loaded solid lipid nanoparticles using response surface methodology. Biomed Micodevices. 2018;20(3):53.

Sasaki H, Sunagawa Y, Takahashi K, Imaizumi A, Fukuda H, Hashimoto T, Wada H, Katanasaka Y, Kakeya H, Fujita M, Hasegawa K, Morimoto T. Innovative preparation of curcumin for improved oral bioavailability. Biol Pharm Bull. 2011;34(5):660–5.

Savarino A, Buonavoglia C, Norelli S, Trani LD, Cassone A. Potential therapies for coronaviruses. Expert Opin Ther Pat. 2006;16(9):1269–88.

Schaffazick SR, Pohlmann AR, De Cordova CAS, Creczkynski Pasa TB, Guterres SS. Protective properties of melatonin-loaded nanoparticles against lipid peroxidation. Int J Pharm. 2005;289:209–13.

Schmidt PJ, Fleming MD. Modulation of hepcidin as therapy for primary and secondary iron overload disorders: preclinical models and approaches. Hematol Oncol Clin North Am. 2014;28(2):387–401.

Schoeman D, Fielding BC. Coronavirus envelope protein: current knowledge. Virol J. 2019;16:69.

Setthacheewakul S, Mahattanadul S, Phadoongsombut N, Pichayakorn W, Wiwattanapatapee R. Development and evaluation of self-microemulsifying liquid and pellet formulations of curcumin, and absorption studies in rats. Eur J Pharm Biopharm. 2010;76(3):475–85.

Shah A, Frost JN, Aaron L, Dronovan K, Drakesmith H. Systemic hypoferremia and severity of hypoxemic respiratory failure in COVID-19. Crit Care. 2020;24:320.

Sheriden C. Convalescent serum lines up as first-choice treatment for coronavirus. Nature Biotechnol 2020;38:655–658. https://doi.org/10.1038/d41587-020-00011-1.

Shetty R, Ghosh A, Honavar SG, Khamar P, Sethu S. Therapeutic opportunities to manage COVID-19/SARS-CoV-2 infection: Present and future. Indian J Ophthalmol. 2020;68(5):693–702.

Shome S, Talukdar AD, Choudhury MD, Bhattacharya MK, Upadhyaya H. Curcumin as potential therapeutic natural product: a nanobiotechnological perspective. J Pharm Pharmacol. 2016;68:1481–500.

Shu T, Huang M, Wu D, Ren Y, Zhang X, Han Y, Mu J, Wang R, Qiu Y, Zhang DY, Zhou X. SARS-coronavirus-2 Nsp13 possesses NTPase and RNA helicase activities that can be inhibited by bismuth salts. Virologica Sinica. 2020;35(3):321–9.

Siddique YH, Khan W, Singh BR, Naqvi AH. Synthesis of alginate-curcumin nanocomposite and its protective role in transgenic Drosophila model of Parkinson's disease. ISRN Pharm. 2013;1–2:794582.

Sillman B, Bade AN, Dash PK. Barghavan B, Kocha T, Mathews S. Creation of a long-acting nanoformulated dolutegravir. Nat Commun. 2018;9:443.

Silva LD, Arrúa EC, Pereira DA, et al. Elucidating the influence of praziquantel nanosuspensions on the in vivo metabolism of Taenia crassiceps cysticerci. Acta Trop. 2016;161:100–5.

Simons K, Toomre D. Lipid rafts and signal transduction. Nat Rev Mol Cell Biol. 2000;1(1):31–9.

Smith KA, Lin X, Bolshakov O, et al. Activation of HIV-1 with nanoparticle-packaged small-molecule protein phosphatase-1-targeting compound. Sci Pharm 2015;83:535–548.

Socha M, Bartecki P, Passirani C, Sapin A, Damge C. Lecompte T. Stealth nanoparticles coated with heparin as peptide or protein carriers. J Drug Targeting. 2009;17(8):575–585.

Sonawane R, Harde H, Katariya M, Agrawal S, Jain S. Solid lipid nanoparticles-loaded topical gel containing combination drugs: an approach to offset psoriasis. Expert Opin Drug Deliv. 2014;11:1833–47.

Song Z, Lu Y, Zhang X, Wang H, Han J, Dong C. Novel curcumin-loaded human serum albumin nanoparticles surface functionalized with folate: characterization and in vitro/vivo evaluation. Drug Des Dev Ther. 2016;10:2643–9.

Sousa F, Castro P, Fonte P, Kennedy PJ, Neves-Petersen MT, Sarmento B. Nanoparticles for the delivery of therapeutic antibodies: dogma or promising strategy? Expert Opin Drug Deliv. 2017;14(10):1163–76.

Srai SK, Chung B, 2Joanne Marks J, 34Katayoun Pourvali K, 2Nita Solanky N, 1Chiara Rapisarda C, Chaston TB, Hanif R, Unwin RJ, Edward S. Debnam ES, Sharp PA. Erythropoietin regulates intestinal iron absorption in a rat model of chronic renal failure. Kidney Int. 2010;78(7):660–667.

Starkloff WJ, Bucala V, Palma SD, et al. Design and in vitro characterization of ivermectin nanocrystals liquid formulation based on a top-down approach. Pharm Dev Technol. 2016;22(6):809–17.

Sui Z, Salto R, Li J, Craik C, De Montellano PRO. Inhibition of the HIV-1 and HIV-2 proteases by curcumin and curcumin boron complexes. Bioorg Med Chem. 1993;1(6):415–22.

Sungnak, W., Huang, N., Bécavin, C. Berg M, Queen R, Litvinukova M et al. SARS-CoV-2 entry factors are highly expressed in nasal epithelial cells together with innate immune genes. Nat Med. 2020;26(5):681–687.

Suravajhala R, Parashar A, Malik B, Nagaraj VA, Padmanaban G, Kavi Kishor P, Polavarapu R, Suravajhala P. Comparative docking studies on curcumin with COVID-19 proteins. Preprints 2020; Preprints 2020; 2020050439. doi: https://doi.org/10.20944/preprints202005.0439.v1.

Swaminathan S, Cavalli R, Trotta F. Cyclodextrin-based nanosponges: a versatile platform for cancer nanotherapeutics development. Nanomed Nanotechnol. 2016;8(4):579–601.

Takada A, Fujioka K, Tsuiji M, Morikawa A, Higashi N, Ebihara H, Kobasa D, Feldmann H, Irimura T, Kawaoka Y. Human macrophage C-type lectin specific for galactose and N-acetylgalactosamine promotes filovirus entry. J Virol. 2004;78(6):2943–2947.

Tanne JH. Covid-19: FDA approves use of convalescent plasma to treat critically ill patients. British Med J 2020;368 m1256. https://doi.org/10.1136/bmj.m1256.

Tavakoli A, Ataei-Pirkooh A, Sadeghi GMM, Bokharaei-Salim F, Sahrapour P, Kiani SJ, Moghoofei M, Farahmand M, Javanmard D, Monavari SH. Polyethylene glycol-coated zinc oxide nanoparticle: an efficient nanoweapon to fight against herpes simplex virus type 1. Nanomedicine. 2018;13(21):2675–90.

Thangapazham RL, Puri A, Tele S, Blumenthal R, Maheshwari RK. Evaluation of a nanotechnology-based carrier for delivery of curcumin in prostate cancer cells. Int J Oncol. 2008;32:1119–24.

Thi TH, Azaroual N, Flament MP. Characterization and in vitro evaluation of the formoterol/cyclodextrin complex for pulmonary administration by nebulization. Eur J Pharm Biopharm. 2009;72:214–8.

Thondawada M, Wadhwani AD, S Palanisamy D, Rathore HS. An effective treatment approach of DPP-IV inhibitor encapsulated polymeric nanoparticles conjugated with anti-CD-4 mAb for type 1 diabetes. Drug Dev Indus Pharm. 2018;44(7):1120–1129.

Tian X, Li C, Huang A, Xia S, Lu S, Shi Z, Lu L, Jiang S, Yang Z, Wu Y, Ying T. Potent binding of 2019 novel coronavirus spike protein by a SARS coronavirus-specific human monoclonal antibody. Emerg Microbes Infect 2020;9(1):382–385. https://doi.org/10.1080/22221751.2020. 1729069.

Trivedi MK, Mondal SC, Gangwar M, Jana S. Immunomodulatory potential of nanocurcumin-based formulation. Inflammopharmacology. 2017;25(6):609–19.

Tsi CJ, Chao Y, Chen CW, Lin WW. Aurintricarboxylic acid protects against cell death caused by lipopolysaccharide in macrophages by decreasing inducible nitric-oxide synthase induction via IkappaB kinase, extracellular signal-regulated kinase, and p38 mitogen-activated protein kinase inhibition. Mol Pharmacol. 2002;62(1):90–101.

Uskokovic V. Nanotechnologies: what we do not know. Technol Soc. 2007;29(1):43–61.

Uskokovic V. Why have nanotechnologies been underutilized in the global uprising against the coronavirus pandemic? Nanomedicine 2020;15(20):1719–1734.

Vajragupta O, Boonchoong P, Watanabe H, Tohda M, Kummasud N, Sumanont Y. Manganese complexes of curcumin and its derivatives: evaluation for the radical scavenging ability and neuroprotective activity. Free Radical Biol Med. 2003;35(12):1632–44.

van der Goot FG, Harder T. Raft membrane domains: from a liquid-ordered membrane phase to a site of pathogen attack. Semin Immunol. 2001;13(2):89–97.

Vankadari N, Wilce JA. Emerging WuHan (COVID-19) coronavirus: glycan shield and structure prediction of spike glycoprotein and its interaction with human CD26. Emerg Microbes Infect. 2020;9(1):601–4.

Vankadari N. Arbidol: a potential antiviral drug for the treatment of SARS-CoV-2 by blocking the trimerization of viral spike glycoprotein? Int J Antimicrob Agents Agents. 2020;56(2):105998.

Velebny S, Hrckova G, Tomasovicova O, Dubinsky P. Treatment of larval toxocarosis in mice with fenbendazole entrapped in neutral and negatively charged liposomes. Helminthologia. 2000;37:119–25.

Verdecchia P, Angeli F, Reboldi G. Angiotensin-converting enzyme inhibitors, angiotensin II receptor blockers and coronavirus. J Hypertens. 2020;38(6):1190–1.

Walia N, Dasgupta N, Ranjan S, Chen L, Ramalingam C. Fish oil based vitamin D nanoencapsulation by ultrasonication and bioaccessibility analysis in simulated gastro-intestinal tract. Ultrason Sonochem. 2017;39:623–35.

Wang M, Hajishengallis G. Lipid raft-dependent uptake, signalling and intracellular fate of Porphyromonas gingivalis in mouse macrophages. Cell Microbiol. 2008;10:2029–42.

Wang M, Cao R, Zhang L, Yang X, Liu J, Xu M, et al. Remdesivir and chloroquine effectively inhibit the recently emerged novel coronavirus (2019-nCoV) in vitro. Cell Res. 2020;30:269–71.

Wang X, Cao R, Zhang H, Liu J, Xu M, Hu H, et al. The anti-influenza virus drug, arbidol is an efficient inhibitor of SARS-CoV-2 in vitro. Cell Discov. 2020;6:28.

Wei P, Fan K, Chen H, Ma L, Huang C, Tan L, Xi D, Li C, Liu Y, Cao A. The N-terminal octapeptide acts as a dimerization inhibitor of SARS coronavirus 3C-like proteinase. Biochem Biophys Res Commun. 2006;339:865–72.

Wen CC, Kuo YH, Jan JT, Liang PH, Wang SY, Liu HG, Lee CK, Chang ST, et al. Specific plant terpenoids and lignoids possess potent antiviral activities against severe acute respiratory syndrome coronavirus. J Med Chem. 2007;50:4087–95.

Wiley JA, Richert LE, Swain SD, Harmsen A, Barnard DL, Randal TD, et al. Inducible bronchus-associated lymphoid tissue elicited by a protein cage nanoparticle enhances protection in mice against diverse respiratory viruses. PLoS ONE. 2009;4(9):e7142.

Wolfram J, Ferrari M. Clinical cancer nanomedicine. Nano Today. 2019;25:85–98.

Wolfram J, Nizzero J, Niu H, Li F, Zhang G, Li Z, Shen H, Blanco E, Ferrari M. Sci Rep. 2017;7:13738.

Wrapp D. De Vlieger KS, Corbett GM, Torres N, Wang W, Van Breedam K, Roose L, van Schie M, Hoffmann S, Pöhlmann, et al. VIB-CMB COVID-19 response team structural basis for potent neutralization of betacoronaviruses by single-domain camelid antibodies. Cell. 2020;181(6):1436–1441.

Wu D, Wu T, Q. Liu Q, Yang Z. The SARS-CoV-2 outbreak: what we know? Int J Infect Dis. 2020;94:44–48.

Xiao C, Dash S, Morgantini C, Patterson BW, Lewis GF. Sitagliptin, a DPP-4 inhibitor, acutely inhibits intestinal lipoprotein particle secretion in healthy humans. Diabetes. 2014;63:2394–401.

Xinyi S. Nanosponges intercept coronavirus infection. Sci Transl Med. 2020;12(550):eabd3078.

Xu S, Cui F, Huang D, Zhang D, Zhu A, Sun X, Cao Y, Ding S, Wang Y, Gao E, Zhang F. PD-L1 monoclonal antibody-conjugated nanoparticles enhance drug delivery level and chemotherapy efficacy in gastric cancer cells. Int J Nanomed. 2018;14:17–32.

Yadav VR, Prasad S, Kannappan R, Ravindran J, Chaturvedi MM, Vaahtera L, Parkkinen J, Aggarwal BB. Cyclodextrin-complexed curcumin exhibits anti-inflammatory and antiprolifera-tive activities superior to those of curcumin through higher cellular uptake. Biochem Pharmacol. 2010;80(7):1021–32.

Yallapu MM, Jaggi M, Chauhan SC. β-cyclodextrin-curcumin self-assembly enhances curcumin delivery in prostate cancer cells. Colloids Surf B. 2010;79:113–125.

Yallapu MM, Jaggi M, Chauhan SC. Curcumin nanoformulations: a future nanomedicine for cancer. Drug Discov Today. 2012;17:71–80.

Yallapu MM, Othman SF, Curtis ET, Bauer NA, Chauhan N, Kumar D. Curcumin-loaded magnetic nanoparticles for breast cancer therapeutics and imaging applications. Int J Nanomed. 2012;7:1761–79.

Yang W, de Villiers MM. Effect of 4-sulphonato-calix[n]arenes and cyclodextrins on the solubi-lization of niclosamide, a poorly water soluble anthelmintic. AAPS J. 2005;7:E241–8.

Yang SNY, Sarah AC, Chunxiao W, Alexander L, Marie BA, Natalie BA, David JA. The broad spec-trum antiviral ivermectin targets the host nuclear transport importin $\alpha/\beta 1$ heterodimer. Antiviral Res. 2020;177:104760.

Yang XX, Li CM, Huang CZ. Curcumin modified silver nanoparticles for highly efficient inhibition of respiratory syncytial virus infection. Nanoscale. 2016;8:3040–8.

Yan G, Yan L, Yucen H, Fengjiang L, Yao Z, Lin C, Tao W, Qianqian S, Zhenhua M, Lianqi Z et al. Structure of the RNA-dependent RNA polymerase from COVID-19 virus. Science. 2020;368:eabb7498.

Yen FL, Wu TH, Tzen CW, Lin LT, Lin CC. Curcumin nanoparticles improve the physicochemical properties of curcumin and effectively enhance its antioxidant and antihepatoma activities. J Agric Food Chem. 2010;58:7376–82.

Yilmaz N, Eren E. Covid-19 and iron gate: the role of transferrin, transferrin receptor and hepcidin. 2020; Presentation: https://www.researchgate.net/publication/340860987.

Yin J, Yasuhiro N, Toshihisa Y. Properties of poly(lactic-co-glycolic acid) nanospheres containing protease inhibitors: camostat mesilate and camostat mesilate. Int J Pharm. 2006;314:46–55.

Zhang L, Pornpattananangkul D, Hu CM, Huang CM. Development of nanoparticles for antimi-crobial drug delivery. Curr Med Chem. 2010;17:585–94.

Zhang Q, Anna Honko A, Zhou J, Gong H, Downs SN, Vasquez JH. Fang RH, Gao W, Griffiths A, Zhang L. Cellular nanosponges inhibit SARS-CoV-2 infectivity. Nano Lett. 2020;20(7):5570–74.

Zhou Y, Jiang X, Tong T, Fang L, Wu Y, Liang J, Xiao S. High antiviral activity of mercap-toethane sulfonate functionalized Te/BSA nanostars against arterivirus and coronavirus. RSC Adv. 2020;10:14161–9.

Zhu N, Zhang D, Wang W, et al.; China novel coronavirus investigating and research team. A novel coronavirus from patients with pneumonia in China, 2019. N Engl J Med. 2020;382(8):727–733.

Zielinska A, Alves H, Marques V, Lurrazo A, Lucarni M, Alves TF, et al. Properties, extraction methods, and delivery systems for curcumin as a natural source of beneficial health effects. Medicina (Kaunas). 2020;56(7):E336.

Zumla A, Chan J, Azhar, EI, Hui DSC, Yuen KYl. Coronaviruses—drug discovery and therapeutic options. Nat Rev Drug Discov. 2016;15:327–347.

Chapter 7
Promising COVID-19 Vaccines

7.1 Vaccine Types

SARS CoV2 vaccines are classified as nucleic acid vaccines, vectored vaccines, recombinant protein vaccines, virus-like particles (VLPs), inactivated vaccines and live-attenuated vaccines (Callaway 2020).

In nucleic acid vaccines, the genetic code (either as DNA or as m RNA) is delivered into the target cell for in situ protein expression. This is very reliable approach in SARS CoV2 vaccine development due to safety, speed, stability and scalability. Moreover, nucleic acid vaccines are advantageous as they elicit CD8$^+$ cytotoxic T cell responses in addition to the normal antibody and CD4$^+$ T cell responses (Pardi et al. 2018; Smith et al. 2020). During vaccine testing, the DNA sequence is conjugated to SARS CoV2 antigenic gene and inserted into the target cell via electroporation and thereafter the gene is expressed as proteins to elicit the immune response. In RNA vaccination, the mRNA encoding a particular SARSs CoV2 antigenic protein usually encapsulated in a lipid coat enables easy entry into the target cell for further expression into immunogenic proteins. Advantages of mRNA vaccines over conventional vaccines include: bypassing the genome integration, the strong/improved immune responses, the rapid development and the generation of multimeric antigens (Pardi et al. 2018).

Vectored vaccines contain a mammalian viral gene as a vector in which the pathogenicity is disabled without altering the replication capacity. The vector is genetically engineered to ligate with the gene encoding the antigenic epitope. The entire recombinant construct will replicate, and the antigenic protein will be expressed to elicit the immune response (Amanat et al. 2020). Spike protein can be expressed as using adenoviral vectors. Adenoviral vectors and chimpanzee adenoviral vectors have been developed for SARS CoV2 vaccines.

Protein vaccination directly uses the antigenic proteins or protein fragments or peptides instead of RNA (Roper et al. 2009). Spike protein, N protein and E protein are used individually or as fusion proteins to function as vaccines against SARS CoV2.

© The Author(s), under exclusive license to Springer Nature Singapore Pte Ltd. 2021   115
Devasena T., *Nanotechnology-COVID-19 Interface*, Nanotheranostics,
https://doi.org/10.1007/978-981-33-6300-7_7

Virus-like particles (VLPs) refer to the external capsid or the shell of the virus which contains only the structural proteins. As VLPs do not contain genetic material, it is not infectious yet can induce immune response (Lu et al. 2010). VLPs of SARS CoV2 can emerge as a good vaccine candidate.

Live-attenuated vaccines are prepared by inactivating the virus and rendering them non-infectious by treatment with chemicals or high temperature. They can replicate but remain non-virulent, and they are mostly intended to produce immunity in single dose. Recently, synthetic genome approaches have been used to generate recombinant SARS CoV2 viruses from genome fragments (Thao et al. 2020). This can be used for the rapid generation of live-attenuated vaccines (Shin et al. 2020).

Inactivated vaccines are SARS CoV2 viral preparations preweakened by mutations to neutralize their pathogenicity. They do not replicate and are much safer than live-attenuated vaccines but require much higher doses due to less immunogenicity. An early vaccine candidate called PiCoVacc capable of protecting monkeys against SARS CoV2 infection was developed (Gao et al. 2020).

Promising vaccine types to fight against SARS CoV2 are summarized in Table 7.1. Like any other vaccine, COVID-19 vaccine also possesses same developmental stage as shown in Table 7.2.

Table 7.1 Possible vaccine candidates for SARS CoV2

Vaccine type	Description	References
DNA vaccine	SARS CoV2 antigen is conjugated to DNA and expressed	Pardi et al. (2018); Smith et al. (2020)
RNA vaccine	RNA encodes the SARS CoV2 antigenic protein	Pardi et al. (2018); Smith et al. (2020)
Vectored vaccine	Adenovirus gene is used as a vector to carry the SARS CoV2 spike protein gene for expression	Amanat et al. (2020)
Protein vaccine	S protein, E protein, M protein or N protein is directly used to induce immune response	Roper et al. (2009)
Virus-like particle (VLP) vaccines	The outer coat of SARS CoV2 devoid of the genetic material is used to elicit immune response	Lu et al. (2010)
Inactivated vaccines	The virus is inactivated by mutation to lose the pathogenicity and retaining the antigenicity	Gao et al. (2020)
Live-attenuated vaccines	The virus is rendered nonpathogenic by chemical or heat treatment without losing the antigenicity	Thao et al. (2020)

Table 7.2 Developmental stages of vaccines

Stage	Name	Description	References
1	Exploratory	Preliminary research investigations on fixing the antigenic part of the target that is capable of eliciting immune response	Singh et al. (2016); Stern et al. (2020)
2	Phase I safety trial (50–100 volunteers)	Checking the immune response, efficacy and tolerance of the vaccine in mammalian models like rats, mice or monkeys	WHO (2013); Stern et al. (2020)
3	Phase II expanded trial (several hundred volunteers)	Checking in a small-size population in order to assess the dosage and the safety profile and confirm the immune response in humans	Villa et al. (2005)
4	Phase III efficacy trial (hundred to thousand volunteers)	The above testing in big size of population	Linhares et al. (2008); Singh et al. (2016)
5	FDA review and approval	Review of trial results by regulatory agency for approval	Shah et al. (2020)
6	Manufacturing and quality control	Large-scale production of approved vaccine and quality checking	Jin et al. (2019)
7	Marketing	Vaccines provided to the public	Ngui et al. (2015)

7.2 Mapping of SARS CoV2-Specific Vaccines

At present, there is no approved vaccine against SARS CoV2. However, premier research organizations such as National Institute of Health and lead biopharmaceutical companies around the world are active under specific stages of development of SARS CoV2 vaccine (Calaway 2020; Paules et al. 2020). In this situation, promising vaccine candidates were proposed, discussed and reviewed by few researchers (Chakraborty 2020; Ahn et al. 2020).

Vaccines are mainly identified by the interaction of specific antigenic epitopes (proteins or peptides) of the target with the host receptor and the strength of the consequent immune response. The S protein is the major target for SARS CoV2 vaccine development. This preference is attributed to the strong interaction of S protein with the ACE2 receptor to mediate attachment and consequent uptake into endosomes for causing infectivity (Kruse et al. 2020). Major histocompatibility complex (MHC)-I and MHC-II epitopes of B cells possess antigenicity against the S protein and behave as a multi-epitopic peptide vaccine (Bhattacharya et al. 2020). A recombinant adenovirus type-5/Ad5-vector vaccine with the spike glycoprotein of a SARS CoV2 strain induces rapid T cell responses from day 14 post-vaccination with a peak at 28th day post-vaccination in healthy adults (Zhu et al. 2020). Tetravalent fusion protein vaccine made out of combinations S protein, RNA-dependent RNA polymerase, E protein and N protein epitopes functions as a multi-epitope vaccine (B and MHC I epitopes). The activity of fusion vaccine was confirmed by docking studies but warrants host expression and further experiments (Tazehkand et al. 2020). The next choice for SARS CoV2 vaccine is the receptor-binding domain of the virus (Jiang et al. 2020). As E protein is involved in crucial stages of the virus life cycle including formation of envelope, pathogenesis, protein assembly and organization, E protein can also be a target for SARS CoV2 vaccine. T cell epitope-based peptide vaccines are capable of targeting SARS CoV2 E protein. MHC class I and MHC class II-binding peptides are also potent vaccine candidates (Abdelmageed et al. 2020).

Alternatively, the ORF-3a and ORF-7a proteins that contribute to the growth and replication of virus can be used as putative T cell epitope determinants. This can result in a prolong CD4 + and CD8 + T cell-mediated immune response. ITLCFTLKR epitope biochemically fitting to HLA allelic proteins was proposed to be a promising vaccine candidate against SARS CoV2 in silico. This epitope has maximum population coverage for different geographical regions (Joshi et al. 2020). HTL, CTL and BCL epitopes of SARS CoV2 proteins are antigenic toward HLA alleles and capable of eliciting humoral and cellular-mediated immune response in silico (Jain et al. 2020).

mRNA-1273 is an mRNA vaccine which encodes S protein of SARS CoV2 and capable of inducing strong immune response in the host. This vaccine was designed by a biopharmaceutical company, Moderna. However, increasing the stability of mRNA, overcoming the innate immune response, enhancement of the cellular uptake efficiency and identification of immune signaling pathways which are specific targets of exogenous mRNA vaccine are challenges to be addressed (Wang et al. 2020).

7.3 Species-Related Vaccines for SARS CoV2

In addition to direct vaccines from SARS CoV2 epitopes, related species such as SARS CoV and MERS CoV can also be used to produce vaccines which can fight SARS CoV2. For example, S protein vaccine obtained from the SARS CoV produces specific neutralizing antibodies against SARS CoV2. Also, CD4[+] T cell responses and CD8[+] T cell responses capable of neutralizing SARS CoV2 are produced by S protein of related species (Martin et al. 2008). CR3022, a neutralizing monoclonal antibody complementary against SARS CoV (ter Meulen et al. 2006; Tian et al. 2020), can interact with the receptor-binding domain of SARS CoV2 S protein. But this vaccine has not been processed beyond phase I trial. Hence, it warrants further study. Subunit vaccines against MERS CoV capable of inducing T cell response and producing high titer of antibodies in vivo may be promising against SARS CoV2 infection (Okba et al. 2017; Bisht et al. 2005).

SARS CoV2 vaccines designed by various companies are under clinical trial stage. The list includes: NVX-CoV2373 (Novavax), PiCoVacc (Sinovac), Ad5-nCoV (CanSino Biologics), AZD1222 (University of Oxford and AstraZeneca), INO-4800 BNT162 (Inovio), (BioNTech and Pfizer), mRNA-1273 (Moderna and NIAID) (Mullard 2020).

SARS CoV2 infection is relatively severe in individuals above 50 years of age for unknown reasons (Amanat et al. 2020). Hence, protecting the population of above 50 years should be an important criterion to be addressed in SARS CoV2 vaccine technology.

7.4 Potent Nanovaccines for SARS CoV2

As no vaccine has successfully been passed all the phases to fight specifically against SARS CoV2, nanomaterials may be availed for effective vaccine design. Advantages of nanocarriers and nanoadjuvants in vaccination against infections have been documented (Malathi et al. 2015; Yue et al. 2012). Integration of nanotechnology concepts to the existing vaccines or the design of novel vaccines using advancements in nanotechnology will deliver a potent vaccine for SARS CoV2 with considerably high immune stimulatory effect. Possible nanovaccines for SARS CoV2 discussed in this section are summarized in Fig. 7.1.

Full-length spike nanoparticles (S-nanoparticles) have been developed as a vaccine for SARS CoV and MERS CoV infection. MERS CoV S-nanoparticles when used with matrix-M1 as adjuvant were capable of preventing MERS CoV infection, suggesting it to be promising vaccines in humans and animals including camels (Coleman et al. 2014, 2017). MERS CoV vaccines fabricated using nanoassembly containing surface proteins such as spike , membrane protein and envelope protein hold potential to raise immunity against SARS CoV2 infection (Kato et al. 2019). Virus-like particles (VLPs) comprising coronavirus proteins and recombinant virus particles are promising vaccine candidate for coronavirus (Coleman et al. 2014). Novavax Company initially produced VLPs containing the S protein fragment and the matrix adjuvant which are immunogenic against MERS CoV infection. In the next stage, Novavax produced nanoparticle vaccines which are immunogenic to the S protein of SARS CoV2 (Liu et al. 2020). Self-assembled α-helical nanopeptides containing the SARS CoV spike protein residues can induce neutralization activity of antisera. This nanopeptide vaccine may be a potent vaccine against SARS CoV2 provided the peptides have confirmed SARS CoV2 spike protein sequence (Pimentel 2009).

SARS CoV DNA vaccine capable of expressing the spike protein encapsulated in polyethylenimine (PEI) nanocarriers for sustained delivery was recently patented. This vaccine induces cellular humoral immune responses and stimulates CD80 and CD86 clusters, class II major histocompatibility complex molecules, IFN-gamma-, TNF-alpha- and IL-2-producing cells (Yoon and Cho 2010). This type of vaccine may have potential to raise the immune response for SARS CoV2 also.

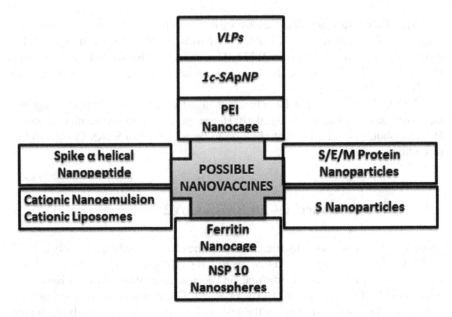

Fig. 7.1 Promising nanovaccines for preventing SARS CoV2 infection (Legend: VLPs—virus-like particles; 1c-SApNP—one-component self-assembling protein nanoparticles; PEI—polyethylene imine; S-nanoparticles—spike protein nanoparticles; S/E/M—spike protein/envelope protein/membrane protein; NSP—non-structural proteins)

SARS CoV or MERS CoV mRNA encoding the spike protein or its fragments when encapsulated into liposomal vesicle functions as an efficient nanovaccine system. This liposomal vaccine stimulates the immune response, elicits high titer of neutralizing antibodies and reduces the viral titer in the lungs of mammalian model (Liu et al. 2020). Thus, the lipid nanoparticles-based vaccines may emerge as potential SARS CoV2 vaccine in the future.

mRNA of MERS CoV structural proteins is capable of inducing immune response when encapsulated in lipid nanoparticles (Yoon and Cho 2010). As the structural proteins of MERS CoV are identical to that of the SARS CoV2, this type of technology can elicit immune response against SARS CoV2 also.

Nsp10 of human SARS CoV2, a 17kD transcription regulatory protein, is used in the development of SARS CoV2 vaccine. Self-assembly is the main strategy involved in this method. The protein self-assembles into nanospheres of 84 angstrom with trigon symmetry which shows unusual surface display of N and the C terminus which exhibits tetrahedral arrangement. The amino and the carboxyl termini fold up to mimic the trimeric receptors which are involved in viral tropism and infection (Carter et al. 2020). This nsp10 vaccine is effective for pulmonary fibrosis, and it can be repurposed for functioning as SARS CoV2 vaccine.

Ufovax, a spin-off vaccine company from Scripps Research, has patented *a self-assembling protein nanoparticle* (1c-SApNP) vaccine technology to fight SARS CoV2. The patent discloses VLPs containing SARS CoV2 spikes protruding from a protein nanoparticle scaffold. 1c-SA*p*NP is capable of inducing immune response and eliciting high titer of neutralizing antibodies against SARS CoV2 (https://www.ufovax.com/).

Cationic liposomes and polymeric nanoparticles are used as carriers for the delivery of DNA vaccines (Lim et al. 2020). Cationic nanoemulsions, liposomes, dendrimers or polysaccharide particles are used in the delivery of SARS CoV2 mRNA vaccines (Pardi et al. 2018 and Zeng et al. 2020). Bacteriophage P22, ferritin nanocages and cowpea mosaic viruses are used as nanoplatforms for SARS CoV2 vaccine delivery (Shin et al. 2020). Promising nanovaccines against SARS CoV2 discussed here are summarized in Fig. 7.1.

7.5 Challenges to Be Addressed During SARS CoV2 Vaccine Development

Antibody-dependent enhancement (ADE) is a post-vaccination phenomenon in which the antibodies induce viral infectivity through Fc receptor pathway. ADE is characterized by an increase in viral entry and viral replication (Tirado et al. 2003). This is an important problem to be addressed while designing SARS CoV2 vaccines. ADE has the theoretical potential to amplify viral infections or trigger immunopathology. In fact, ADE has already been demonstrated during vaccine development against Dengue, Zika viruses and Ebola virus (Khandia et al. 2018; Katzelnick et al. 2017; Bardina et al. 2017; Kuzmina et al. 2018). Now, ADE is associated with Betacoronavirus and SARS CoV2 pandemic also (Wang et al. 2014; Tetro 2020; Wan et al. 2020). Clinical findings, immunologic assays, biomarkers, in vitro studies and in vivo models are not reliable to predict ADE because the protective and the detrimental antibody responses cannot easily be differentiated. Therefore, undoubtedly, ADE remains as a main challenge to be addressed during SARS CoV2 vaccine development. This insists the need for safety assessment in humans as the vaccine hunt is constantly increasing for SARS CoV2 (Arvin et al. 2020).

The problem of ADE is less if the non-surface proteins are used for vaccine design. For instance, vaccines targeting N protein of SARS CoV2 may challenge the ADE and prevent the viral entry and replication after the administration of vaccines (Kim et al. 2004).

Next to ADE, mutation frequency is another challenge in vaccine technology. SARS CoV2 shows greater mutation frequency (refer Chap. 1) as compared to DNA viruses which is attributed to the variations in the genes coding for the RBD of the spike glycoprotein. Hence, mutation factor should also be favorably addressed while developing vaccine for COVID-19 (Zhou et al. 2020).

Overall, integration of the ADE and the gene mutation factors with the recent advancements in nanotechnology, SARS CoV vaccines and MERS CoV vaccines may help in the repurposing of SARS CoV2 vaccine. However, the proposed and patented vaccine candidates should be validated in different phases for marketing and safe usage by the public.

7.6 Biopharmaceutical Companies Are Active in Constructing SARS CoV2 Vaccines

After the SARS CoV2 pandemic, various companies are actively involved in the production of vaccines for COVID-19 and they are in different phases of clinical trials. The vaccines are expected to come into the market shortly. The details of the vaccines under trial with different companies are enlisted in Table 7.3. Codagenix, a clinical-stage synthetic biology company (New York), uses software to recode the genomes of viruses for designing live-attenuated vaccines. In collaboration with serum institute of India, Codagenix has developed a live-attenuated vaccine for SARS CoV2 named as CDX-005. This vaccine is expected to mimic wild-type SARS CoV2 antigen. CanSino biological in China genetically engineered an adenovirus vector to express the spike glycoprotein of SARS CoV2 in humans. The company has also successfully published the peer-reviewed data on the assessment of safety and immunogenicity of the product. Innovio is a biotechnology company involved in the production of DNA-based products to manage life-threatening diseases. Innovio has developed a DNA vaccine candidate called "INO4800" which is under final phase of development. INO4800 is capable of generating a balanced antibody and T cell immune responses which is very important in the case of SARS CoV2 vaccine. Entos Pharmaceuticals, in collaboration with Precision NanoSystems Inc., has developed a DNA vaccine for COVID-19 using the proprietary *Fusogenix* drug delivery system. Fusogenix is a proteo-lipid vehicle (PLV) formulation that uses a novel mechanism

Table 7.3 Pharmaceutical companies involved in the SARS CoV2 vaccine development

Company	Location	SARS CoV2 vaccine type	References
Codagenix	New York	Live-attenuated vaccine for SARS CoV2	https://codagenix.com
CanSino Biologics	China	Adenovirus type 5 vectored vaccine (Ad5-nCoV) for SARS CoV2	Zhu et al. (2020)
Inovio Pharmaceuticals	San Diego	DNA vaccine	Zhu et al. (2020)
Entos Pharmaceuticals Inc. and Precision NanoSystems Inc	Canada	DNA vaccine	https://www.entospharma.com https://www.precisionnanosystems.com
Moderna	USA	mRNA vaccine	Zhu et al. (2020)
BioNTech-Pfizer	Germany	mRNA vaccine	Mulligan et al. (2020)

of action to deliver active molecules directly into the cytosol of the target cells. Moderna has developed its mRNA-1273 vaccine candidate for SARS CoV2. The mRNA encodes the S-2P antigen of the SARS CoV2 spike glycoprotein. The vaccine is formulated as a lipid nanocapsule containing a fixed ratio of mRNA and lipid capable of inducing immune response against SARS CoV2 (Jackson et al. 2020). BioNTech and Pfizer have developed an mRNA vaccine candidate "BNT162b1" for COVID-19. BNT162b1 is capable of eliciting strong $CD4^+$ and $CD8^+$ T cell responses against SARS CoV2-receptor-binding domain.

References

Abdelmageed MI, Abdelmoneim AH, Mustafa MI , Elfadol NM, Murshed NS, Shantier SW, Makhawi AM. Design of a multiepitope-based peptide vaccine against the E protein of human COVID-19: an immunoinformatics approach. BioMed Res Int. 2020; 2683286.

Ahn DG, Shin HY, Kim MH, Lee S, Kim HS, Myoung J, Kim BT, Kim SJ. Current status of epidemiology, diagnosis, therapeutics, and vaccines for novel coronavirus disease 2019 (COVID-19). J Microbiol Biotechnol. 2020; 30(3):313–24.

Amanat F, Krammer F. SARS-CoV-2 vaccines: status report. Immunity. 2020;52(4):583–9.

Arvin AM, Fink K, Schmid MA, Cathcaert A, Spreafico M, Havenar-Doughton C, et al. A perspective on potential antibody-dependent enhancement of SARS-CoV-2. Nature. 2020. https://doi.org/10.1038/s41586-020-2538-8.

Bardina SV, Bunduc P, Tripathi S, Duehr J, Frere JJ, Brown JA, Nachbagauer R, et al. Enhancement of Zika virus pathogenesis by preexisting antiflavivirus immunity. Science. 2017;356:175–80.

Bhattacharya M, Sharma AR, Patra P, Ghosh P, Sharma G, Patra BC, et al. Development of epitope-based peptide vaccine against novel coronavirus 2019 (SARS-COV-2): immunoinformatics approach. J Med Virol. 2020;92(6):618–31.

Bisht H, Roberts A, Vogel L, Subbarao K, Moss B. Neutralizing antibody and protective immunity to sars coronavirus infection of mice induced by a soluble Recombinant polypeptide containing an n-terminal segment of the spike glycoprotein. Virology. 2005;334:160–5.

Callaway E. The race for coronavirus vaccines: a graphical guide: eight ways in which scientists hope to provide immunity to SARS-CoV-2. Nature. 2020;580:576–7.

Carter DC, Wright B, Gray Jerome W, Rose JP, Wilson E. A unique protein self-assembling nanoparticle with significant advantages in vaccine development and production. J Nanomater. 2020; 4297937.

Chakraborty C, Sharma AR, Sharma G, Bhattacharya M, Lee SS. Sars-CoV-2 causing pneumonia-associated respiratory disorder (COVID-19): diagnostic and proposed therapeutic options. Eur Rev Med Pharmacol Sci. 2020;24:4016–26.

Coleman CM, Liu YV, Mu H, Taylor JK, Massare M, Flyer DC, Smith GE, Frieman MB. Purified coronavirus spike protein nanoparticles induce coronavirus neutralizing antibodies in mice. Vaccine. 2014;32(26):3169–74.

Coleman CM, Venkataraman T, Liu YV, Glen GM, Smith GE, Flyer DC, Frieman MB. MERS-CoV spike nanoparticles protect mice from MERS-CoV infection. Vaccine. 2017;35(12):1586–9.

Gao Q, Bao L, Mao H, Wang L, Xu K, Yang M, et al. Rapid development of an inactivated vaccine candidate for SARS-CoV-2. Science. 2020;369:77–81.

Jackson LA, Anderson EJ, Rouphael NG, Roberts PC, Makhene M, Coler RN, et al. An mRNA Vaccine against SARS-CoV-2—preliminary report. New Engl J Med. 2020. https://doi.org/10.1056/NEJMoa2022483.

Jain N, Shankar U, Majee P, Kumar A. Scrutinizing the SARS-CoV-2 protein information for the designing an effective vaccine encompassing both the T-cell and B-cell epitopes. 2020; https://doi.org/10.1101/2020.03.26.009209.

Jiang S, Du L, Shi Z. An emerging coronavirus causing pneumonia outbreak in Wuhan, China: calling for developing therapeutic and prophylactic strategies. Emerg Microbes Infect. 2020;9:275–7.

Jin J, Tarrant RD, Bolam EJ et al. Production, quality control, stability, and potency of cGMP-produced plasmodium falciparum RH5.1 protein vaccine expressed in Drosophila S2 cells. npj Vaccines 2018; 3:32.

Joshi A, Joshi BC, Amin-ul Mannan M, Kaushik V. Epitope based vaccine prediction for SARS-COV-2 by deploying immuno-informatics approach. Inform Med Unlocked. 2020;19:100338.

Kato T, Takami Y, Kumar Deo V, Park EY. Preparation of virus-like particle mimetic nanovesicles displaying the S protein of middle east respiratory syndrome coronavirus using insect cells. J Biotechnol. 2019;306:177–84.

Katzelnick LC, Gresh L, Halloran ME, Mercado JC, Kuang G, Gorcon A, Balmseda A, Haris E. Antibody-dependent enhancement of severe dengue disease in humans. Science. 2017; 358:929–32.

Khandia R, Munjal A, Dhama K, Karthik K, Tiwari R, Malik YS, et al. Modulation of Dengue/Zika virus pathogenicity by antibody-dependent enhancement and strategies to protect against enhancement in Zika virus infection. Front Immunol. 2018;9:597.

Kim TW, Lee JH, Hung CF, Peng S, Roden R, Wang MC, et al. Generation and characterization of DNA vaccines targeting the nucleocapsid protein of severe acute respiratory syndrome coronavirus. J Virol. 2004;78:4638–45.

Kruse RL. Therapeutic strategies in an outbreak scenario to treat the novel coronavirus originating in Wuhan, China. 2020; 9:72.

Kuzmina NA, Younan P, Glichuk P, Ramanathan P, et al. Antibody-dependent enhancement of Ebola virus infection by human antibodies isolated from survivors. Cell Reports. 2018;24(1802–1815):e5.

Lim M. Md Badruddoza AZ, Firdous J, Azad M, Adnan Mannan A, Ahmed Al-Hilal T et al. Engineered nanodelivery systems to improve DNA vaccine technologies. Pharmaceutics. 2020; 12:30.

Linhares AC, Velázquez FR, Pérez-Schael I, Sáez-Llorens X, Abate H, Espinoza F et al. Human rotavirus vaccine study group. Efficacy and safety of an oral live attenuated human rotavirus vaccine against rotavirus gastroenteritis during the first 2 years of life in Latin American infants: a randomised, double-blind, placebo-controlled phase III study. Lancet. 2008; 371:1181–9.

Liu C, Zhou Q, Li Y, Garner LV, Watkins SP, Carter LJ, Smoot J, Gregg AC, Daniels AD, Jervey S. Research and development on therapeutic agents and vaccines for COVID-19 and related human coronavirus diseases. ACS Central Sci. 2020;6:315–31.

Lu B, Huang Y, Huang L, Li B, Zheng Z, Chen Z, Chen J, Hu Q, Wang H. Effect of mucosal and systemic immunization with virus-like particles of severe acute respiratory syndrome coronavirus in mice. Immunology. 2010;130(2):254–61.

Malathi B, Mona S, Devasena T, Kaliraj P. Immunopotentiating nano-chitosan as potent vaccine carter for efficacious prophylaxis of filarial antigens. Int J Biol Macromol. 2015;73:131–7.

Martin JE, Louder MK , Holman LA , Gordon IJ, Enama ME , Larkin BD , Andrews CA, Vogel L, Koup RA , Roederer M, Bailer RT , Gomez PL, Nason M, Mascola JR, Nabel GJ, Graham BS. VRC 301 study team. A SARS DNA vaccine induces neutralizing antibody and cellular immune responses in healthy adults in a phase I clinical trial. Vaccine 2008; 26:6338–6343.

Mullard A. COVID-19 vaccine development pipeline gears up. Lancet. 2020;395:1751–2.

Mulligan MJ, Lyke KE, Kitchin N et al. Phase I/II study of COVID-19 RNA vaccine BNT162b1 in adults. Nature 2020; 586, 589–93. https://doi.org/10.1038/s41586-020-2639-4.

Ngui EM, Hamilton C, Nugent M, Simpson P, Willis E. Evaluation of a social marketing campaign to increase awareness of immunizations for urban low-income children. World Mycotoxin J. 2015;114(1):10–5.

Okba NM, Raj VS, Haagmans Bl. Middle east respiratory syndrome coronavirus vaccines: current status and novel approaches. Current Opin Virol. 2017; 23:49–58.

Pardi N, Hogan MJ, Porter FW, Weissman D. mRNA vaccines—a new era in vaccinology. Nat Rev Drug Discov. 2018;17:261–79.

Paules CI, Marston HD, Fauci AS. Coronavirus infections—more than just the common cold. JAMA. 2020;323(8):707–8.

Pimentel TA, Yan Z, Jeffers SA, Holmes KV, Hodges RS, Burkhard P. Peptide nanoparticles as novel immunogens: design and analysis of a prototypic severe acute respiratory syndrome vaccine. Chem Biol Drug Des. 2009;73(1):53–61.

Roper RL, Rehm KE. SARS vaccines: where are we? Expert Rev Vaccines. 2009;8:887–98.

Shah A, Marks PW, Hahn SM. Unwavering regulatory safeguards for COVID-19 vaccines. JAMA. 2020;324(10):931–2.

Shin MD, Shukla S, Chung YH Beiss V, Khim Chan S, Ortega-Rivera OA et al. COVID-19 vaccine development and a potential nanomaterial path forward. Nature Nanotechnol 2020; 15:646–55.

Singh K, Mehta S. The clinical development process for a novel preventive vaccine: an overview. J Postgrad Med. 2016;62(1):4–11.

Stern PL. Key steps in vaccine development. Anals Allergy Asthma Immunol. 2020;125(1):17–27.

Smith TRF, Patel A, Ramos S, Elwood D, Zhu X, Yan J, et al. Immunogenicity of a DNA vaccine candidate for COVID-19. Nature Commun. 2020;11:2601.

Tazehkand MN, Hajipour O. Evaluating the vaccine potential of a tetravalent fusion protein against coronavirus (COVID-19). J Vaccines Vaccin. 2020;11:412. https://doi.org/10.35248/2157-7560.20.11.412.

Ter Meulen J, van den Brink EN, Poon LLM, Marissen WE, Leung CSW, Cox F, et al. Human monoclonal antibody combination against SARS coronavirus: synergy and coverage of escape mutants. PLoS Med. 2006;3(7):e237.

Tetro JA. Microbes and infection. Is COVID-19 receiving ADE from other coronaviruses? 2020; 22:72–3.

Thao, TTN, Labroussaa F, Ebart N, V'kovski P, Stalder H, Prottman J et al. Rapid reconstruction of SARS-CoV-2 using a synthetic genomics platform. Nature. 2020; 582:561–5.

Tian X, Li C, Huang A, Xia S, Lu S, Shi Z, Lu L, Jiang S, Yang Z, Wu Y, Ying T. Potent binding of 2019 novel coronavirus spike protein by a SARS coronavirus-specific human monoclonal antibody. Emerg Microbes Snf Infect. 2020;9:382–5.

Tirado SM, Yoon KJ. Antibody-dependent enhancement of virus infection and disease. Viral Immunol. 2003;16:69–86.

Villa LL, Costa RL, Petta CA, Andrade RP, Ault KA, Giuliano AR, et al. Prophylactic quadrivalent human papillomavirus (types 6, 11, 16, and 18) L1 virus-like particle vaccine in young women: A randomised double-blind placebo-controlled multicenter phase II efficacy trial. Lancet Oncol. 2005;6:271–8.

Wan Y, Shang J, Sun S, Tai W, Chen J, Geng Q, et al. Molecular mechanism for antibody-dependent enhancement of coronavirus entry. J Virol. 2020;94(5):e02015-e2019.

Wang F, Kream RM, Stefano GB. An evidence based perspective on mRNA-SARS-CoV-2 vaccine development. Med Sci Monitor. 2020; 26:e924700-1–8.

Wang SF, Tseng SP, Yen CH, Yang JY, Tsao CH, Shen CW, et al. Antibody-dependent SARS coronavirus infection is mediated by antibodies against spike proteins. Biochem Biophys Res Commun. 2014;451:208–14.

World Health Organization. Causality assessment of an adverse event following immunization (AEFI). User manual for the revised WHO classification. World Health Organization. 2013; 1–43.

Yoon CH, Cho JS. SARS vaccine nano-delivery system. 2010; Patent No: KR20100120473A.

Yue H, Wei W, Fan B Yue Z, Wang L, Ma G, Su Z. The orchestration of cellular and humoral responses is facilitated by divergent intracellular antigen trafficking in nanoparticle-based therapeutic vaccine. Pharmacol Res. 2012; 65(2):189–97.

Zeng C, Hou X, Yan J, Zhang C, Li W, Zhao W et al. Leveraging mRNAs sequences to express SARS-CoV-2 antigens in vivo. 2020; https://doi.org/10.1101/2020.04.01.019877.

Zhou P, Yang XL, Wang XG, Hu B, Zhang L, Zhang W, Si HR, Zhu Y, Li B, Huang CL, Chen HD, Chen J, Luo Y, Guo H, Jiang RD, Liu MQ, Chen Y, Shen XR, Wang X, Zheng XS, Zhao K, Chen QJ, Deng F, Liu LL, Yan B, Zhan FX, Wang YY, Xiao GF, Shi ZL. A pneumonia outbreak associated with a new coronavirus of probable bat origin. Nature. 2020;579:270–3.

Zhu FC, Li YH, Guan XH, Hou LH, Wang WJ, Li JX. Safety, tolerability, and immunogenicity of a recombinant adenovirus type-5 vectored COVID-19 vaccine: a dose-escalation, open-label, non-randomised, first-in-human trial. Lancet. 2020;395:1845–54.

Websites

https://synbiobeta.com/codagenix-announces-the-synthesis-and-preliminary-safety-of-scalable-live-attenuated-vaccine-candidate-against-covid-19/.

https://cen.acs.org/pharmaceuticals/vaccines/CanSino-publishes-first-COVID-19/98/i21.

https://ir.inovio.com/news-releases/news-releases-details/2020/INOVIO-Announces-Positive-Int erim-Phase-1-Data-For-INO-4800-Vaccine-for-COVID-19/default.aspx.

https://www.entospharma.com/news/entos-pharmaceuticals-partners-with-precision-nanosy stems-to-enable-highly-scalable-gmp-manufacturing-of-fusogenix-dna-covid-19-vaccine.

https://www.pfizer.com/news/press-release/press-release-detail/pfizer-and-biontech-announce-early-positive-update-german.

https://www.entospharma.com.

https://www.precisionnanosystems.com; https://codagenix.com.

Chapter 8
Viral Assays to Detect the Effects of Nanoparticles on SARS CoV2

8.1 Uses of Viral Assays in COVID-19 Management

Viral assays are analytical and imaging techniques used for evaluating various factors which would help in understanding the (i) structure of virus and host cells, (ii) interaction points between virus and host, viral entry, (iii) infection mechanisms, (iv) fate of nanoparticulate drugs and vaccines, (v) cytotoxicity and (vi) immune response. Relating the aspects of COVID-19 and SARS CoV2 using these techniques is summarized in Table 8.1. Overall, these assays are useful in understanding different aspects of SARS CoV2 which includes:

- Determining the antiviral activity of nanoparticles by computing the plaque-forming units per milliliter
- Influence of therapeutic nanoparticles on viral titer
- High-resolution imaging of cellular uptake of nanovaccines
- Detection of live cell population to confirm the safety of the nanoparticulate drugs
- Monitor the infected host cells
- Elucidating the cell fusion process
- Illustrating the glycoprotein-mediated infection mechanism
- Imaging of drug–RNA complex and drug–virus complex
- Structural modules of the virus
- Guest–host interactions
- Vaccine targets and therapeutic targets
- Antibody titer raised by the infection
- Detection of SARS CoV2 neutralizing antibodies
- Effect of nanoparticles on different stages of viral life cycle.

© The Author(s), under exclusive license to Springer Nature Singapore Pte Ltd. 2021   127
Devasena T., *Nanotechnology-COVID-19 Interface*, Nanotheranostics,
https://doi.org/10.1007/978-981-33-6300-7_8

Table 8.1 Techniques used for determining different aspects of SARS CoV2

Technique	Aspects	References
Viral plaque assay	Measurement of viral titer and its correlation with infection Investigating the antiviral activity of functionalized nanoparticles	Medhi et al. (2020), Rogers et al. (2008), Rodriguez-Izquierdo et al. (2020)
TCID50 assay	To determine the 50% infectious dose and the effect of antivirals To determine the production of SARS CoV2 antibodies by nanoencapsulated drugs	Ghaffari et al. (2019), Wu et al. (2020)
Confocal virion adsorption assay	To determine the course of viral infection and the virus-induced changes in the host cell in the presence and absence of functional nanoparticles	Medhi et al. (2020)
β-galactosidase assay	To monitor the mode of antiviral action of functional nanoparticles To determine the viral life cycle stages which are the targets for the nanoparticles For investigating vaccine expression in the host cells	Merten et al. (2006), Bar et al. (2006), York et al. (2018)
Transmission electron microscopy and cryo-electron microscopy	To elucidate the structural features with good resolution for identifying anti-SARS CoV2 targets Targeting ion channels involved in SARS CoV2 infection	Ma et al. (2020), Kern et al. (2020)
Western blotting	To analyze SAR CoV2 proteins targeted by antivirals To analyze anti-SARS CoV2 antibody titer	Ou et al. (2020), Tai et al. (2020)
Flow cytometry	To understand the immunological consequences of SARS CoV2 infection To identify promising targets for the design and discovery of SARS CoV2 vaccines and drugs	Cossarizza et al. (2020)
In silico approaches (molecular docking)	To predict virus–host interaction, drug–viral protein interaction, vaccine–protein interaction	Padilla-Sanchez (2020), Peele et al. (2020)

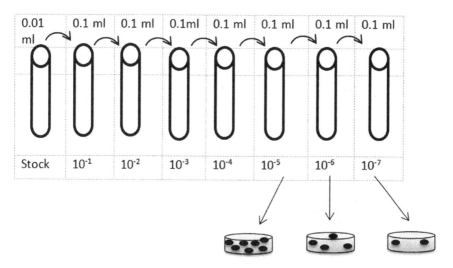

Fig. 8.1 Viral plaque assay

8.2   Viral Plaque Assay

One of the most significant viral assays is the viral plaque assay (Fig. 8.1). It is the way of viral titer measurement which correlates with the infection. The antiviral activity of nanomaterial can very well be evaluated using the plaque assay. In this assay, the viral stock is diluted tenfold and 0.1 ml aliquots are layered on to cell monolayers and incubated for cellular attachment. The monolayer is then covered with an agar nutrient medium. The infected cells release the progeny which is marked by circular zones called plaque, which can be stained by crystal violet for good visualization. The antiviral activity of nanoparticles can be accurately found by computing the plaque-forming units per milliliter (PFU/mL) in the presence and absence of the nanoparticles. For example, antiviral activity of silver nanoparticles and gold nanoparticles and their inhibitory concentrations can be determined by plaque assay (Rogers et al 2008; Rodriguez-Izquierdo et al 2020).

8.3   The TCID50 Assay

The TCID50 (median tissue culture infectious dose) is a method for measuring viral titer. TCID50 refers to the viral concentration required to infect 50% of the monolayered cells. This method allows rapid quantification of viral titer in the presence and absence of nanoparticles. Serially diluted viral stock is incubated with cell monolayer and allowed for infection. Cells can be infected with a fivefold dilution series of coronavirus containing media in triplicate and cultured for a further 48 h. Culture plates can be fixed in 4% paraformaldehyde for 5 min at room temperature and then stained with 0.05% crystal violet in 20% methanol for 30 min at room temperature. After rinsing the plates with water, live cells can be detected as evidenced by the presence of the stain. Viral load was calculated based on the dilution at which the media killed 50% of the cells using the TCID50 formula. TCID50 is used to quantify infectious units in cell culture and to determine the infectious dose. Various biotechnology companies are marketing TCID50 kits for specific viruses. TCID50 assay can be used for evaluating the antiviral activity of functional nanoparticles. For example, antiviral activity of zinc oxide nanoparticles, PEGylated zinc oxide nanoparticles and quercetin nanoparticles against H1N1 influenza virus and avian virus has, respectively, been evaluated by this assay (Ghaffari et al 2019). This concept can be applied for SARS CoV2 assay also. In this context, SARS CoV spike protein nanoparticles were used to find out the degree of infection in vaccine development studies. The production of neutralizing antibodies was assessed by TCID50 assay (Coleman et al 2014).

8.4   Confocal Virion Adsorption Assay

In this method, a confocal microscope is used to determine the course of viral infection and the virus-induced changes in the host cell in the presence and absence of functional nanoparticles preferably using immunofluorescence (Chen and Liang 2020). Uptake of nanoparticles used for any applications such as antiviral and vaccine can be imaged using confocal microscopy. Confocal laser scanning microscopy (CLSM) is a contemporary high-resolution optical technique that can be used to derive comprehensive image of the cellular structure with high spatial resolution in the presence of contrast agents. Gold nanoparticles can be used to enhance contrast as it is free of disadvantages like photobleaching and possesses high stability (Lemelle et al 2008).

8.5   The β-Galactosidase Entry Assay

The β-galactosidase entry assay is an FDA approved protocol for evaluating the virus entry and detecting the effect of vaccines on host. β-galactosidase hydrolyzes the glycosidic bonds in the disaccharide lactose to form galactose and glucose, thus providing energy and carbon source. In virology, β-galactosidase is used as a reporter gene to monitor the viral entry and infection in the host cell. The capacity of functional nanoparticles used for enhancing the antiviral activities of existing drugs can be monitored by analyzing the infected host cells. Galactosidase is used to monitor the mode of antiviral action of functional nanoparticles in specific and also to determine the viral life cycle stages which are the targets for the nanoparticles. For instance, mechanistic insights into the antiviral activity of silver nanoparticles have been elucidated in HIV strains using β-galactosidase assay (Lara et al. 2010). As silver nanoparticles have many potential applications in the management of COVID-19, β-galactosidase assay is highly significant in the assay stages of COVID-19 research. Infection can be assessed after two days of incubation by quantifying the activity of the β-galactosidase produced after infection. The 50% inhibitory concentration (IC50) can be defined according to the percentage of infectivity inhibition relative to the positive control. β-Galactosidase indicator gene is also used in cell fusion assay in order to identify novel antivirals. In this assay, cells expressing viral coat glycoproteins are fused with the cells expressing the entry receptors and the β-galactosidase indicator gene is used to imitate coated virus–cell fusions. This assay provides insights into virus–cell fusion processes and enables us to gain more perceptions into the glycoprotein-mediated infection mechanism (Merten et al 2006; Bar et al 2006; York et al. 2018). As spike glycoprotein is one of the infection mechanisms of SARS CoV2, galactosidase assay deserves significance.

8.6 TEM Micrographing

Transmission electron and transmission electron cryo-micrographs can help us to visualize the structure of viruses and the infected cells and demonstrate the viricidal ability of functional nanoparticles and also for characterization of the effects of functional nanoparticles on virus-infected cells. In fact, the three-dimensional atomic map of SARS CoV2 has been created using cryo-electron microscope and the nanosize of SARS CoV2 was also revealed initially using cryo-EM ultrastructure map (Gui et al. 2017). TEM images have disclosed the nature of binding of drug with SARS CoV2 and the RNA–drug complex structure in order to elucidate the antiviral mechanism of the drug with reference to blocking of SARS CoV2 RNA replication (Ma et al. 2020). This type of ultra-imaging is possible even if nanoparticles or nanoparticles-bound drugs are used in virology studies.

A very significant use of TEM was achieved in the study of ion channels of SARS CoV involved in the host infection. SARS CoV2 encodes three putative ion channels such as E, 8a and 3a. In SARS CoV, 3a is involved in viral release, inflammasome activation and cell death. Thus, 3a can be targeted for treating COVID-19. This notion however needs a thorough study of the structure and interaction of 3a which essentially needs TEM imaging. Cryo-TEM micrograph reveals a novel fold in 3a in Betacoronaviruses that infect bats and humans, opening a new avenue for targeting SARS CoV2 (Kern et al. 2020; David et al. 2020).

8.7 Western Blotting

Western blot assay is a technique used to assay proteins and protein modifications using preprocessing followed by gel electrophoresis. In the context of coronavirus, the structure of SARS CoV2 is made of proteins, and the antibodies elicited against the infection are also proteins. Some classes of antiviral agents target the proteins and modify them during therapeutic interventions. Hence, Western blotting assay may play a major role in COVID-19 research. As far as COVID-19 is concerned, Western blot assay is used for the study of SARS CoV2 proteins and their interaction with antivirals. Western blot assay can be used for the detection of SARS coronavirus-positive sera using N195 protein as an antigen (He et al. 2020). Cellular entry and the immune cross-reactivity of SARS CoV2 glycoprotein can be elucidated based on Western blotting (Ou et al. 2020).

Western blot analysis can interpret the receptor-binding domain (RBD) of SARS CoV2. Thereafter, the domain has been suggested as potential viral attachment inhibitor site and this confirms the possibility for the development of RBD-based vaccines for fighting against COVID-19 (Tai et al. 2020).

A truncated 195-amino-acid fragment from the C terminus of the nucleocapsid protein (N195) has ability to detect SARS CoV2 antibodies without cross-reactivity. Hence, Western blot assay involving N195 can be used to analyze SARS CoV2-positive sera with high specificity and sensitivity. Thus, this assay is very significant in diagnosis of COVID-19 (He et al. 2004a, b).

8.8   Flow Cytometry Assay

The basic principle of flow cytometry involves the passage of fluorescently labeled cells in single file in front of a laser, in order to excite them. Thereafter, the emitted wavelength is analyzed in order to elucidate the physical and chemical features of different cell populations and also to enable cell sorting. Flow cytometry is a multiparameter analyzing strategy which can be used to understand the immunological consequences of SARS CoV2 infection, to identify promising targets for the design and discovery of SARS CoV2 vaccines and to screen antiviral molecules. When using flow cytometer for virology, the virus structures such as the capsid and the viral glycoproteins can be tagged with fluorescent tags like green fluorescent proteins or other fluorophores. Antibodies directed against the virus surface glycoproteins such as spike proteins can be coupled to nanoparticles for greater detectability by light scattering. The RNA viral genome can be labeled with several different dyes (Syto13, Syto62, SyBr green-I, YOYO-1, TOTO-1 and PicoGreen) for detection (McSharry et al. 1994). Biological and immunological features of the coronavirus infection can be illustrated using the features of T and B lymphocytes and other inflammatory cells with the help of flow cytometry (Prompetchara et al. 2020). Flow cytometry is used as an important analytical technique to disclose the details of immune response against SRAS CoV2 which will be useful in repurposing of vaccine and antiviral targets (Cossarizza et al. 2020). Discriminating viruses, study of single virus genomics, vaccine quality control, antiviral design, nucleocapsid sorting and surface protein analysis can be done for SARS CoV2 with the help of flow cytometry. A strategic progress is the combination of flow cytometry and mass spectrometry into an improvised technique called as mass cytometry or cytometry-by-time-of-flight (CyTOF) (Bandura et al. 2009). This innovative technique relies on tagging of samples with antibodies coupled to rare metal isotopes rather than fluorescent moieties, where abundance is measured by mass spectrometry. CyTOF lacks spectral overlaps and irrelevance of background auto-fluorescence (Nowicka et al. 2017).

Following are the potential applications of flow cytometry which can be used in COVID-19 management:

- Detection and profiling of pro-inflammatory and innate cytokines which are elevated in the blood of SARS CoV2-infected patients.
- Rapid identification of human immune cell subsets such as monocytes, neutrophils, eosinophils, T cells, B cells and NK cells with high reproducibility.
- Sterile sorting of infectious materials.
- Complete immunophenotyping of T cell subsets and other effector cells involved in the viral immune response. Immunophenotyping is an important procedure to test the viability of repurposed vaccines in animal model.
- High-throughput detection of receptor-binding and receptor expression profiling.
- Illustrating the functions of innate immunity in terms of monocytes, macrophages, dendritic cells, natural killer cells and their capacity in governing the early phases of the infection.
- Analyzing the production and utilization of compounds involved in cytokine storm such as cytokines, chemokines and their receptors.
- Disclosing the kinetics of neutralizing antibodies produced as a result of the humoral response to viral infection.
- To elucidate the role of B cells, plasma cells, T cells, CD4$^+$ and CD8$^+$ clusters against viral infection.
- Understanding the viral antigenic epitopes and the corresponding specific T cell response.
- Interpreting immune activation and immune suppression in COVID-19.

8.9 Real-Time Polymerase Chain Reaction Assay (RT-PCR Assay)

In the traditional PCR method, the gene is amplified and the PCR products or the amplicon is run on the agarose gel or PAGE to analyze the amplification. But in the real-time PCR (RT-PCR), the amplification during each PCR cycle is monitored in a real-time manner and quantified primarily using the fluorescent dye or using the fluorescent labeled oligonucleotides. The viral genes expressed in the host cells need to be amplified in order to confirm the viral infection. In addition, the amplification is also used to indicate the host cell status following the antiviral therapy targeting the genome of the virus. Hence, RT-PCR can also be used as an indicator for the antiviral ability of drug and nanoencapsulated drug candidates. Use of optically tunable nanoparticles in RT-PCR-based detection of SARS CoV is detailed under Chap. 5 of this book.

8.10   *In Silico* Approaches: Simulation and Docking

In computer simulation, models of natural systems like the proteins or genomes of the viruses and host cells are created and the outcome of virus–host interaction, the effect of drug–protein interaction, vaccine–protein interaction, etc., can be predicted. Various software and docking models are being used for the computational design of protein inhibitors. Molecular docking and dynamic simulations for antiviral compounds against SARS CoV2 have been carried out by several workers (Peele et al. 2020). Phylogenetic analysis of SARS CoV2 genome, molecular docking and molecular dynamics simulations are some of the in silico approaches used in the study of repurposed antiviral drugs to bind in the active site of SARS CoV2 main protease, the important target for COVID-19 control. Cresset Flare software is used for molecular docking studies against the spike protein SARS CoV2 (PDB ID: 6VSB). PkCSM and SwissADME Web servers are used for in silico prediction studies on the pharmacokinetics properties and the safety profile of the potential drug candidates (Kirana et al. 2020). AutoDock Vina software is used for the prediction of drug–enzyme active site docking (Kumar et al. 2020). In silico analysis of SARS CoV2 spike glycoprotein and insights into antibody binding and visualization can be enabled using molecular visualization and analysis software such as: UCSF Chimera and Rosetta Dock and other bioinformatics tools including SWISS-MODEL, Bridges Large, Stampede and Frontera (Padilla-Sanchez and Peele et al. 2020).

References

Bandura DR, Baranov VI, Ornatsky OI, Antonov A, Kinach R, Lou X, Pavlov S, Vorobiev S, Dick JE, Tanner SD. Mass cytometry: technique for real time single cell multitarget immunoassay based on inductively coupled plasma time-of-flight mass spectrometry. Anal Chem. 2009;81:6813–22.

Bar S, Takada A, Kawaoka Y, Alizon M. Detection of cell-cell fusion mediated by Ebola virus glycoproteins. J Virol. 2006;80(6):2815–22.

Chen L, Liang J. An overview of functional nanoparticles as novel emerging antiviral therapeutic agents. Mater Sci Eng C. 2020;112:110924.

Coleman CM, Liu YV, Mu H, Taylor JK, Massare M, Flyer DC, Smith GE, Frieman MB. Purified coronavirus spike protein nanoparticles induce coronavirus neutralizing antibodies in mice. Vaccine. 2014;32(26):3169–74.

Cossarizza A, De Biasi S, Guaraldi G, Girardis M, Mussini C. SARS-CoV-2, the virus that causes COVID-19: cytometry and the new challenge for global health. Cytometry A. 2020;97(4):340–3.

Ghaffari H, Tavakoli A, Moradi A, Tabarraei A, Salim FB, Zahmatkeshan M, et al. Inhibition of H1N1 influenza virus infection by zinc oxide nanoparticles: another emerging application of nanomedicine. J Biomed Sci. 2019;26:70.

Gui M, Liu X, Guo D, Zhang Z, Yin CC, Chen Y, Xiang Y. Electron microscopy studies of the coronavirus ribonucleoprotein complex. Protein Cell. 2017;8(3):219–24.

He Q, Chong KH, Chng HH, et al. Development of a Western blot assay for detection of antibodies against coronavirus causing severe acute respiratory syndrome. Clin Diagn Lab Immunol. 2004;11(2):417–22.

He Q, Chong KH, Chng HH, Leung B, Ling AE, Wei T, Chan SW, Ooi EE, Kwang J. Development of a Western blot assay for detection of antibodies against coronavirus causing severe acute respiratory syndrome. Clin Diagn Lab Immunol. 2004;11(2):417–22.

Kern DM, Sorum B, Hoel CM, Sridharan S, Remis J, Toso DB, Brohawn SG. Cryo-EM structure of the SARS-CoV-2 3a ion channel in lipid nanodiscs. 2020; https://doi.org/10.1101/2020.06.17. 156554.

Kiran G, Karthik L, Shree Devi MS, Sathiyarajeswaran P, Kanakavalli K, Kumar KM, Ramesh Kumar D. In Silico computational screening of Kabasura Kudineer—official siddha formulation and JACOM against SARS-CoV-2 spike protein. J Ayurveda Integr Med 2020 (In Press).

Kumarab Y, Singhc H, Patel CN. In silico prediction of potential inhibitors for the Main protease of SARS-CoV-2 using molecular docking and dynamics simulation based drug-repurposing. J Infect Publ Health. 2020 (In Press).

Lara HH, Ayala-Nunez V, Turrent I, Liliana, Rodríguez-Padilla C. Mode of antiviral action of silver nanoparticles aginst HIV-1. J Nanobiotechnol 2010; 8:1–10.

Lemelle A, Veksler B, Kozhevnikov IS, Akchurin GG, Piletsky SA, Meglinski I. Application of gold nanoparticles as contrast agents in confocal laser scanning microscopy. Laser Phys Lett. 2008;6(1):71–5.

Ma Q, Pan W, Li R, Liu B, Li C, Xie Y, et al. Liu Shen capsule shows antiviral and anti-inflammatory abilities against novel coronavirus SARS-CoV-2 via suppression of NF-κB signaling pathway. Pharmacol Res. 2020;158:104850.

McSharry JJ. Uses of flow cytometry in virology. Clin Microbiol Rev. 1994;7(4):576–604.

Medhi R, Srinoi P, Ngo N, Tran HV, Lee TR. Nanoparticle-based strategies to combat COVID-19. ACS Appl Nano Mater. 2020;3(9):8557–80.

Merten CA, Stitz J, Braun G, Medvedovska J, Cichutek K, Buchholz CJ. Fusoselect: cell–cell fusion activity engineered by directed evolution of a retroviral glycoprotein. Nucleic Acids Res. 2006;34(5):e41.

Nowicka M, Krieg C, Weber LM, Hartmann FJ, Guglietta S, Becher B, Levesque MP, Robinson MD. CyTOF workflow: differential discovery in high-throughput high-dimensional cytometry datasets. F1000 Res. 2017; 6:748.

Ou X, Liu Y, Lei X, Li P, Mi D, Ren L, et al. Characterization of spike glycoprotein of SARS-CoV-2 on virus entry and its immune cross-reactivity with SARS-CoV. Nature Commun. 2020;11(1):1620.

Padilla-Sanchez. In silico analysis of SARS-CoV-2 spike glycoprotein and insights into antibody binding. Res Ideas Outcomes. 2020; 6:e55281.

Peele KA, Chandrasai P, Srihansa T, Krupanidhi S, Sai Ayyagari V, Babu DJ et al. Molecular docking and dynamic simulations for antiviral compounds against SARS-CoV-2: a computational study. Inform Med Unlock 2020; 19:100345.

Prompetchara E, Ketloy C, Palaga T. Immune responses in COVID-19 and potential vaccines: lessons learned from SARS and MERS epidemic. Asia Pacific J Allergy Immunol. 2020;38(1):1–9.

Rodriguez-Izquierdo I, Serramia MJ, Gomez R, De La Mata FJ, Bullido MJ, Muñoz-Fernández MA. Gold nanoparticles crossing blood-brain barrier prevent HSV-1 infection and reduce herpes associated amyloid-β secretion. J Clin Med. 2020; 9(1):155.

Rogers JV, Parkinson CV, Choi YW, Speshock JL, Hussain SM. A preliminary assessment of silver nanoparticle inhibition of monkeypox virus plaque formation. Nanoscale Res Lett. 2008;3(4):129–33.

Tai W, He L, Zhang X, Pu J, Voeonin J, Jiang S. Characterization of the receptor-binding domain (RBD) of 2019 novel coronavirus: implication for development of RBD protein as a viral attachment inhibitor and vaccine. Cell Mol Immunol. 2020;17:613–20.

Wu K, Saha R, Su D, Krishna VD, Liu J, Cheeran M, Wang JP. Magnetic-nanosensor-based virus and pathogen detection strategies before and during COVID-19. ACS Appl Nano Mater. 2020;3(10):9560–80.

York J, Nunberg JH. A Cell-cell fusion assay to assess arenavirus envelope glycoprotein membrane-fusion activity. Methods Mol Biol. 2018;1604:157–67.

Chapter 9
Non-pharmaceutical Interventions for COVID-19 Management

At the moment, the therapeutic strategies to deal with the infection are only supportive; prevention aimed at reducing transmission in the community is our best weapon. Non-pharmaceutical interventions (Fig. 9.1) including mask wearing, hand hygiene and social distancing in response to COVID-19 which seems to be a major epidemic-prevention factor (Noh et al. 2020) are discussed in this chapter.

9.1 Masking

The two ways of virus transmission are large droplets and small aerosol particles. More likely greater the transmission by larger droplets, higher the preventive effects of hand hygiene. That is, hand hygiene may not be constructive if viruses are transmitted as small aerosol particles. There is evidence that ten times greater quantities of viruses are required for infection via intranasal drops than with aerosols (Alford et al. 1966). There are evidences for transmission of SARS CoV2 as droplets. Highly sensitive laser light scattering studies on speech droplets generated by asymptomatic carriers of severe acute respiratory syndrome coronavirus 2 (SARS CoV2) reveal that thousands of fluid droplets are emitted per second. The droplets are of 12- to 21-μm size prior to dehydration and 4 μm after dehydration (Stadnytskyi et al. 2020). The diameters of these particles are in the micron range. These particles are too small to settle because of gravity, but they are carried by air currents and dispersed by diffusion and air turbulence (Meselson 2020). These observations not only recommend the use of protective aids but also reveal that there is a substantial probability that normal speaking causes airborne virus transmission in confined environments.

A hot debate is on pertaining to the use of face masks (including cloth and surgical) as a prevention tool in the community vis-à-vis the recent recommendations by the World Health Organization. To shed light on this important topic, Esposito et al. reviewed relevant literature related to use of mask in combating the transmission of

Fig. 9.1
Non-pharmaceutical
interventions to prevent
COVID-19 transmission
(Legend: MIH—mask
wearing plus instant hand
hygiene approach;
PPE—personal protective
equipment)

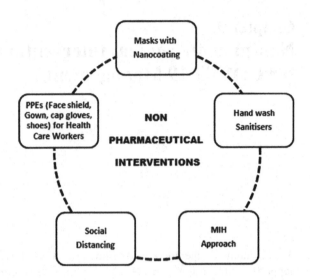

SARS CoV. They recommend that cloth masks are a simple, economic and sustainable alternative to surgical mask as a means of source control of SARS CoV2 for general community (Esposito et al. 2020). Mask use may provide a reduction in respiratory illness regardless of hand washing practices (Aiello et al. 2010). Hence, a high degree of compliance will get the best out of the universal masking in public.

Nanomaterials-based masks can help to capture the aerosol, kill the viruses and help in preventing the transmission of COVID-19. Silver nanoparticle-based filters are highly efficient in killing aerosolized viruses (Joe et al. 2016). Silicon-based nanoporous materials can be used as a hard mask during the pattern-transfer process onto a flexible lightweight polymeric thin film in an RIE system. The intrinsic hydrophobic nature of the membrane attributes to antifouling and self-cleaning property (El-Atab et al. 2020). Electrospun nanofibers coated with TiO_2 and deposited on a filter surface can capture submicrometer droplets. This can destroy the virus upon UV irradiation or under natural sunlight (Lee et al. 2010 and Konda 2020).

Coatings of graphene could be exploited with nanoparticles of silver nitrate and titanium dioxide for better trapping of pathogens (Li et al. 2006). Copper oxide incorporated into face mask textiles exhibits antiviral activity (Borkow et al. 2010). Graphene oxide films are also useful for creating protective, breathable barrier layers in fabric masks as they contribute hydrophobic and dry surface (Steinberg et al. 2017). Protective graphene face masks could be recycled by photocatalysis or heat. Mild heat treatment at 56 °C for 30 min can denature the viruses adsorbed in the masks (Darnell et al. 2002). Hence, graphene may have a leading role in the fight against COVID-19 (Palmieri et al. 2020).

Two-dimensional carbides and nitrides (MXenes) are worth exploring as anti-SARS CoV2 nanocoatings owing to their unique properties like hydrophilicity, photocatalytic capacity, biocompatibility and stability. Ti_3C_2Tx MXene interacts with amisno acids and enables binding and immobilization of viral spike glycoprotein and consequent deactivation of virus. Photocatalytic property can be used to degrade the adsorbed viruses in the presence of light (Weiss et al. 2020). Copper salt nanoparticles (chloride, iodide, sulfide, etc.), which are known for having an antiviral effect, could be helpful in the development of PPEs like nonwoven overshoes, surgical gowns, hair cups, respirators, etc., with improved shielding properties. This could help in preventing the unwanted nosocomial virus spreading by medical personnel (Sportelli et al. 2020).

It is known that ACE2 receptors of the host cell are entry gate for SARS CoV2. The possibility of functionalizing ACE2 on to the surface of quantum dots and nanoflowers has already been proposed. The nanoparticles when coated to precursors of PPEs like masks, clothes, gloves, etc., will be able to capture SARS CoV2 with high affinity, thereby preventing their host cell entry (Aydemir et al. 2020).

9.2 Hand Hygiene

Hand hygiene is of course a fair defensible measure as SARS CoV, MERS CoV, influenza virus and closely related human coronavirus, remained infectious and survived on surface materials common to public and domestic areas for several days (Otter et al. 2016; Warnes et al. 2015). WHO recommends that good hand hygiene mediated by hand wash or sanitizer is one of the most basic, yet powerful strategies to reduce the spread of COVID-19. One of the disadvantages of frequent hand washing or sanitizing is the skin surface damage (manifested eczema, maceration and erosion) which may be a portal entry for viruses (Lan et al. 2020). Still, application of hand cream/moisturizers on intact skin after hand washing or proper use of protective hand gloves can prevent skin damage during preventive strategies against COVID-19 (Yan et al. 2020). A city-based start-up (Weinnovate Bio solutions) has developed a marketable non-alcoholic and non-inflammable sanitizing solution incorporated with silver nanoparticles under DST funding. Nanomaterial-based sanitizer is expected to prevent the synthesis of viral molecules and help disinfecting surfaces of skin and environmental stuffs. Silver nanoparticles owing to their size have the ability to preclude the genesis of viral negative-strand RNA and viral budding.

9.3 The MIH Approach

Ma et al. used avian influenza virus (AIV) to mock SARS CoV2 because they are both enveloped and pleomorphic spherical viruses with a diameter of around 80–120 nm. Thereafter, they evaluated the efficacy of: three types of masks in blocking avian influenza virus (AIV) in aerosols and instant hand wiping in removing AIV from hands. The findings disclosed that instant hand wiping using a wet towel soaked in water containing 1.00% soap powder, 0.05% active chlorine or 0.25% active chlorine from sodium hypochlorite removed 98.36%, 96.62% and 99.98% of the virus from hands, respectively. N95 masks, medical masks and homemade masks made of four-layer kitchen paper and one-layer cloth could block 99.98%, 97.14% and 95.15% of the virus in aerosols. Analyzing the results, the MIH approach (i.e., mask-wearing plus instant hand hygiene approach) was proposed to slow the exponential spread of the virus. This MIH approach has been supported by the experiences of seven countries in fighting against COVID-19 (Ma et al. 2020).

9.4   Social Distancing

Social distancing is another non-pharmaceutical intervention for COVID-19. Scientific and ethical basis for social distancing interventions against COVID-19 has very recently been discussed by Lewnard and Lo (Lewnard and Lo 2020). During the 2003 SARS CoV outbreak in Singapore, numerous non-pharmaceutical interventions were implemented successfully for disease control. The potential effect of social distancing interventions on SARS CoV2 spread and COVID-19 burden in Singapore and recommend that social distancing could substantially reduce the number of SARS CoV2 infections. (Koo et al. 2020). In China, aggressive social distancing implemented by closure of schools, workplaces, religious buildings, roads and transit systems; cancelation of public gatherings; mandatory quarantine of uninfected people without known exposure to SARS CoV2; and large-scale electronic surveillance has helped to fight the COVID-19 and has led to a progressive reduction of cases (Kupferschmidt et al. 2020). In Hong Kong, 44% reduction in influenza transmission in the community was estimated in early February 2020, after the implementation of social distancing to control the spread of COVID-19 (Cowling et al. 2020). Although these actions have been praised by WHO, the possibility of imposing similar measures in other countries raises important questions. Social distancing measures with established guidelines should be implemented to reduce disease burden, even if SARS CoV2 vaccines are developed in the future. In particular, social distancing will contribute to control infections in vaccine mismatch years (Noh et al. 2020).

References

Aiello AE, Murray GF, Perez V, et al. Mask use, hand hygiene, and seasonal influenza-like illness among young adults: a randomized intervention trial. J Infect Dis. 2010;201(4):491–8.

Alford RH, Kasel JA, Gerone PJ, Knight V. Human influenza resulting from aerosol inhalation. Proc Soc Exp Biol Med. 1966;122(3):800–4.

Aydemir D, Ulusu NN. Angiotensin-converting enzyme 2 coated nanoparticles containing respiratory masks, chewing gums and nasal filters may be used for protection against COVID-19 infection. Travel Med Infect Dis. 2020; 101697.

Borkow G, Zhou SS, Page T, Gabbay J. A novel anti-influenza copper oxide containing respiratory face mask. PLoS ONE. 2010;5(6):1–9.

Cowling BJ, Ali ST, Ng TW, Tsang TK, Li JC, Fong MW, et al. Impact assessment of non-pharmaceutical interventions against coronavirus disease 2019 and influenza in Hong Kong: an observational study. Lancet Public Health. 2020;5(5):e279–88.

Darnell MER, Subbarao K, Feinstone SM, Taylor DR. Inactivation of the coronavirus that induces severe acute respiratory syndrome SARS-CoV. J Virol Methods. 2004;121:85–91.

El-Atab N, Qaiser N, Badghaish H, Shaikh SF, Mustafa Hussain M. Flexible nanoporous template for the design and development of reusable anti-COVID-19 hydrophobic face masks. ACS Nano. 2020; 14(6):7659–65.

Esposito S, Principi N, Leung CC, Migliori GB. Universal use of face masks for success against COVID-19: evidence and implications for prevention policies. Eur Respiratory J 2020 (In press).

Joe YH, Park DH, Hwang J. Evaluation of Ag nanoparticle coated air filter against aerosolized virus: anti-viral efficiency with dust loading. J Hazard Mater. 2016;301:547–53.

Konda A, Prakash A, Moss GA, Schmold M, Grant GD, Guha S. Aerosol filtration efficiency of common fabrics used in respiratory cloth masks. ACS Nano. 2020;14:6339.

Koo JR, Cook AR, Park M, Sun Y, Sun H, Lim JT, et al. Interventions to mitigate early spread of COVID-19 in Singapore: a modelling study. Lancet Infect Dis. 2020;20(6):678–88.

Kupferschmidt K, Cohen J. China's aggressive measures have slowed the coronavirus. They may not work in other countries. 2020; https://www.sciencemag.org/news/2020/03/china-s-aggressive-measures-have-slowed-coronavirus-they-may-not-work-other-countries.

Lan J, Song Z, Miao X. Skin damage among healthcare workers managing coronavirus disease-2019. J Am Acad Dermatol. 2020;82(5):1215–6.

Lee BY, Behler K, Kurtoglu ME, Wynosky-Dolfi MA, Rest RF, Gogotsi Y. Titanium dioxide-coated nanofibers for advanced filters. J Nanopart Res. 2010;12:2511–9.

Lewnard JA, Lo NC. Scientific and ethical basis for social-distancing interventions against COVID-19. Lancet Infect Dis. 2020;20(6):631–3.

Li Y, Leung P, Yao L, Song QW, Newton E. Antimicrobial effect of surgical masks coated with nanoparticles. J Hosp Infect. 2006;62:58–63.

Ma QX, Shan H, Zhang HL, Li GM, Yang RM, Chen JM. Potential utilities of mask wearing and instant hand hygiene for fighting SARS-CoV-2. J Med Virol 2020; 1–5.

Meselson M. Droplets and aerosols in the transmission of SARS-CoV-2. New Engl J Med 2020; 382:2063

Noh JY, Seong H, Yoon JG, Song JY, Cheong HJ, Kim WJ. Social distancing against COVID-19: implication for the control of influenza. J Korean Med Sci. 2020;35(19):e182.

Otter JA, Donskey C, Yezli S, Douthwaite S, Goldenberg SD, Weber DJ. Transmission of SARS and MERS coronaviruses and influenza virus in healthcare settings: the possible role of dry surface contamination. J Hosp Infect. 2016;92(3):235–50.

Palmieri V, Papi M. Can graphene take part in the fight against COVID-19? Nano Today. 2020;33:100883.

Sportelli MC, Izzi M, Kukushkina EA, Hossain SI, Picca RA, Ditaranto N, Cioffi N. Can nanotechnology and materials science help the fight against SARS-CoV-2? Nanomaterials. 2020;10(4):802.

Stadnytskyi V, Bax CE, Bax A, Anfinrud P. The airborne lifetime of small speech droplets and their potential importance in SARS-CoV-2 transmission. Proc Natl Acad Sci USA. 2020;117(22):2006874117.

Steinberg SR, Cruz M, Mahfouz NGA, Qiu Y, Hurt RH. Breathable vapor toxicant barriers based on multilayer graphene oxide. ACS Nano. 2017;11:5670–9.

Warnes SL, Little ZR, Keevil CW. Human coronavirus 229E remains infectious on common touch surface materials. mBio. 2015; 6(6):e01697–15.

Weiss C, Carriere M, Fusco L, Capua I, Regla Nava JA, Pasquali M et al. Toward nanotechnology-enabled approaches against the COVID-19 pandemic. ACS Nano. 2020; 14(6): 6383–406.

Yan Y, Chen H, Chen L. Consensus of Chinese experts on protection of skin and mucous membrane barrier for healthcare workers fighting against coronavirus disease 2019. Dermatol Therapy. 2020; e13310.

Websites

https://www.who.int/southeastasia/news/detail/04-05-2020-promote-hand-hygiene-to-save-lives-
and-combat-covid-19.
https://timesofindia.indiatimes.com/city/pune/silver-nanoparticles-infused-sanitizer-on-the-anvil/
articleshowprint/75089018.cms.

Printed in the United States
By Bookmasters

Printed in the United States
by Baker & Taylor Publisher Services